2.6 Administrative Division

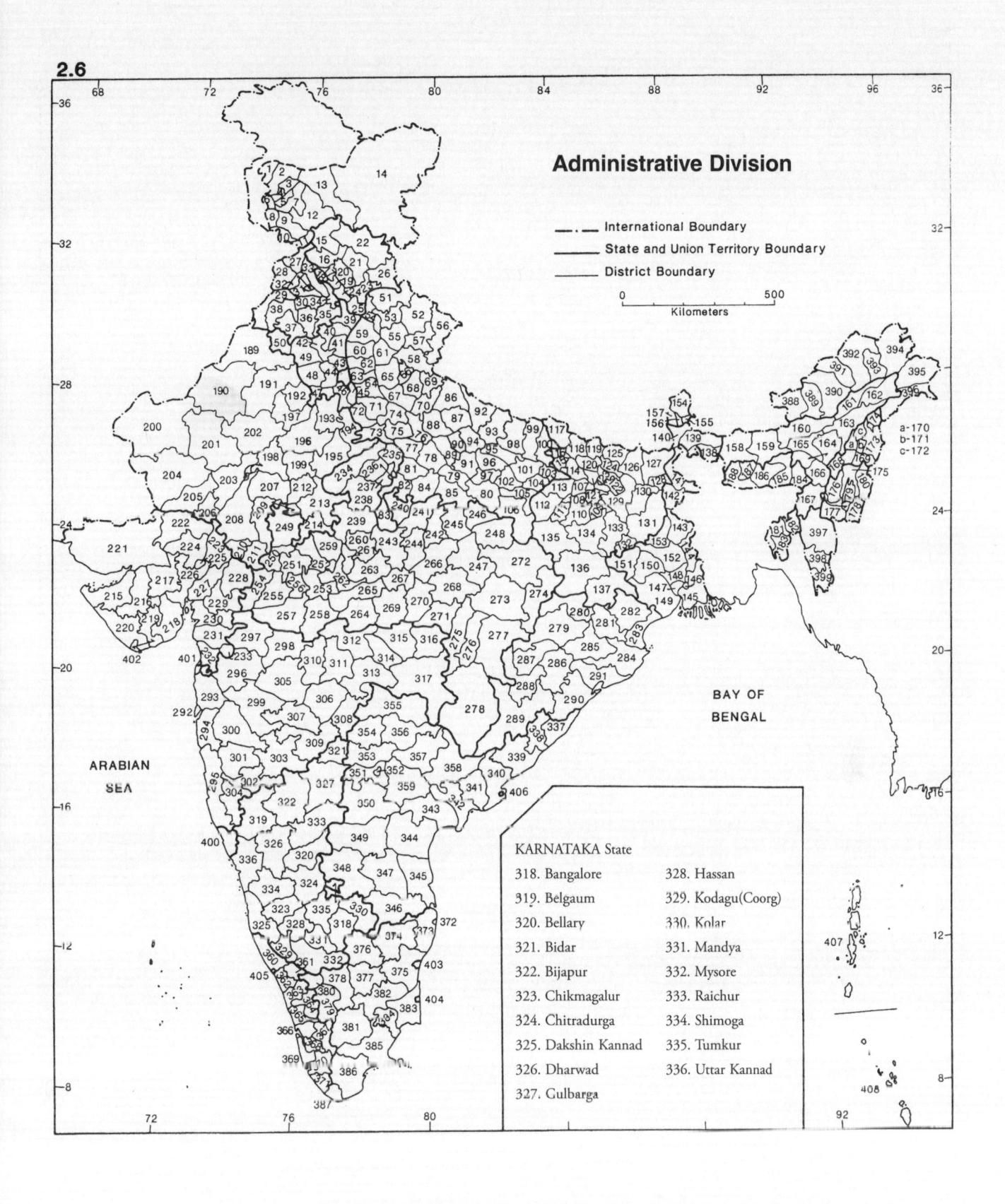

- — · — International Boundary
- ——— State and Union Territory Boundary
- ——— District Boundary

0 — 500 Kilometers

ARABIAN SEA

BAY OF BENGAL

KARNATAKA State

318. Bangalore	328. Hassan
319. Belgaum	329. Kodagu (Coorg)
320. Bellary	330. Kolar
321. Bidar	331. Mandya
322. Bijapur	332. Mysore
323. Chikmagalur	333. Raichur
324. Chitradurga	334. Shimoga
325. Dakshin Kannad	335. Tumkur
326. Dharwad	336. Uttar Kannad
327. Gulbarga	

Regional Dynamics in Modernizing India

Regional Dynamics in Modernizing India

An Experience in Western Ghats Region

Iwao MAIDA

Senshu University Press

Regional Dynamics in Modernizing India
An Experience in Western Ghats Region

Book design/typography by Hiromi Kyoo
Printed and bound at the Fujiwara-Printing Co., Ltd.

© 2010 by Iwao Maida
Published by Senshu University Press, 2010
3-8 Kanda-Jinbocho, Chiyoda-ku, Tokyo, 101-0055, Japan
ISBN 978-4-88125-233-8

Tungabahdra Dam ; 49m H, Length 2,450m, command area covers some 500,000ha

Settlement landscape Bidarakere, situated drought prone area

Malabar Coast facing Arabian Sea near Goa

Catholic church amidst hillside of Western Ghats Mountain range

Goa, One of the most oldest Catholic church in India

Goa, former Portuguese colony (1510 ~ 1961) one of the strongholds of Indian Subcontinent for Propagating Christianity and trade fortress

Matrimonial ceremony led by Brahmin priest, Yerdona

Infant marriage ceremony

Social stratification and widening economic gap ; residence of prosperous rich farmer Yerdona

Shabby huts constructed by in-migrants and seasonal workers, *coolie* Yerdona

Mechanization underway among the innovative rich farmers Yerdona

Irrigation rush and construction rush Yerdona

Changing means of transportation, Portuguese-styled two wheeler, *chariot* Yerdona

Private owned community bus Yerdona

Small boy, weeping alone just in front of the thatched house while parents are working out daylong

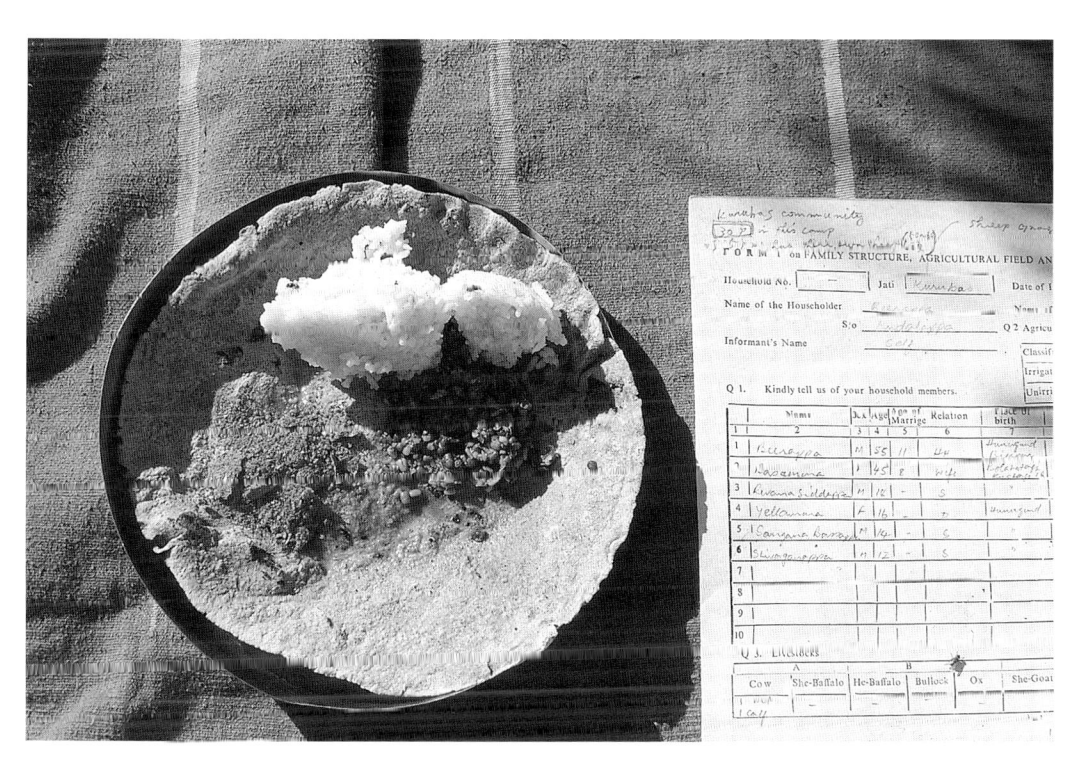

In-depth focused interviewing by applying questionnaire

Robust and sturdy jeep for traverse and reconnaissance survey, Goa in front of Portuguese styled with balcony

CONTENTS

PROLOGUE

CHAPTER 1 **RECENT TRENDS IN THE SOCIO-ECONOMIC STRUCTURE OF VILLAGES IN THE CENTRAL KARNATAKA** 1

Introduction 1

1. Population in Bidarakere 2
    1.1 Growth of Population 2
    1.2 Structure of Population 3
    1.3 Structural Changes in Population 5

2. Family Structure 5
    2.1 Family Composition 6
    2.2 Family Types 7

3. Inheritance System and Family Types 9

4. Recent Trends in Agricultural Management and Its Perfomance 17
    4.1 Agro-climatic conditions; an overview 18
    4.2 Cropping Patterns in Bidarakere 19
    4.3 Comparative Analysis of Crop-combination Types in Bidarakere and Yerdona 20
    4.4 Productivities of Major Crops in Bidarakere and Yerdona 29

5. Farming Size and Economic Performances of Farm Management 38

6. A New Frontier and "Irrigation Rush" in Yerdona 44

CHAPTER 2 **POPULATION AND OCCUPATIONAL STRUCTURE OF MALNAD VILLAGES** 51

1. Population and Occupational Structure 51

2. Distribution of Household by *Jati* 55

CHAPTER 3 **RECENT TRENDS IN RICE CULTIVATION AND PLANTATION ECONOMY OF THE WESTERN GHATS** 59

1. Macroscopic Overview 59

2. Recent Trends in Agricultural Land Use 60

3. Agricultural Production 62
    3.1 Paddy 62
    3.2 Plantation Crops 63
        3.2.1 Coffee 63
            Brief History of Coffee Plantation 64
            Agricultural Calendar of Coffee Plantation 65
            Economic Performance 66
        3.2.2 Cardamom 67
            Environmental Factors of Cardamom Plantation 67
        3.2.3 Gross Incomes 68

CHAPTER 4 **AGRICULTURAL LAND UTILIZATION AND PRODUCTIVITIES IN NARAVI** 71

1. Agricultural Land Utilization 71

2. Productivities and Circulation of Agricultural Reproduction 71

3. Horticultural Crops 82

CHAPTER 5 **KARNATAKA LAND REFORM ACT AND ITS IMPACTS ON THE SOCIO-ECONOMIC CONDITIONS OF NARAVI** 85

1. Preface 85

2. Land Reforms in Naravi; an Overview 85

3. Landlordism in Naravi before the Land Reforms Act 86

4. Legislative Aspects of Radical Land Reforms 87

5. Land Reforms in Naravi and its Critical Appraisal 90

EPILOGUE

INDEX

PROLOGUE

India was once often referred as poverty-stricken country, and, still now some of intellectual in western world frequently tends to point out that she has long been caught by the powerful clutch of vicious circle of poverty. In many respects up to the present time, they say much yet remains to be done. Apples are likely to be yet scarcer among the have-nots. Admittedly, as far as I know, when I was staying in India, came across actual rural scenes. There must be undeniably some due reasons that traditionally held images among the western intellectuals toward India were supposed to be the fact. But, at the same time, I thought it. Was it really a some of plausible parts of real world they live. *viz* a fragmentary part of actual reality.?

Various fancies flitted through my mind. When I came down to the localities, which were selected out as a sample survey villiges by Research Project Team of Hiroshima University, After examined, some of them, turned out to be true, the other clealy false.

So I felt it better to start at once to thoroughly scrutinize a true picture of multi-facetted aspects of modernizing India over time minutely, from geographical point of views.

My perspectives are concerned widely with research topics ranging from physiographic, agro-climatic, socio-economic and to also socio-cultural dimensions as well.

So the research plan was scrutinized, and then reshuffled to set up once again in the form of topic wise *geomatrix* in an unified way.

In this context, main focuses were summarized and placed on the following categorized research topics; Geographical field research work was carried out in accordance with this concept.

Main objectives are thus set up as follows;
1) to depict family lineage over generations and relationships between inheritance and *dowry* system.
2) to follow up landholding and tenure system over generations.
3) to typify and map a various types of irrigation systems ranging from either distributary channels, tank well or tube well in each regions.
4) to classify major types of farming systems and find out typal structure of crop combinations and measure up its productivities in each sample survey villages.
5) to describe precisely a changing settlement patterns in historical perspective.
6) to make clear a series of dynamic processes of social stratification over time. In the sample survey villages, by applying In-depth focused interviewing and questionnaire.

Author has tried to follow up a salient features in rural scenes, to make clear a series of rural dynamics in Western Ghats region.

More than one quarter years has passed since I, myself participated as a member of Geographical Research Project Team of Hiroshima University in India.

It took too much time for me to make critical appraisal in order to trace properly up the salient features in structural changes in India, with special reference to an experience in Western Gahts region over the last half century.

Hisotry speaks what has happened. That is why I have undertaken a pioneering geographical research work. I will give testimony that what has been accutually taking place in modernizing India since mid 20 century.

I will expect hopefully this research monograph will serve to describe what has really happened in the regional geography of South India and to give plausible answer to explain the time when and the reason why India has narrowly succeeded in take-off among emerging countries in the third world, groaning under the extraordinary population pressure next to China

Increasing degree of self-sufficiency of food stuffs in India has been cleary revealed and well reflected on a remarkable development in agricultural sector in the last half a century.

It should be born in mind the fact that Indian economy has shown remarkable progress in pararell with

Geographical profiles of sample survey villages in the research projects

Field survey		The first, 1978		The second, 1980		The third, 1982	
Topographical division		South *Maidan*		North *Maidan*		*Malnad* and West Coast	
Climatic condition		Wet savanna with annual rain of 700 to 800mm		Dry savanna and semi-arid area with annual rainfall less than 600mm		Tropical monsoon and rain forests with annual rainfall more than 1500mm	
Soils		Deep red soil		Shallow red soil and black cotton soil		Red soil and lateritic soils	
Land use and irrigation		Wet land by tank & well irrigations and dry land	Wet land by canal irrigation	Dry land by rain-fed cultivation	Change from dry land to wet land by canal irrigation	Paddyfield and plantation	Paddyfield by rain-fed and channel irrigation
Development programme	Large-scale development by canal irrigation	Aralamallige (M, III, m)	Chikkamaralli (M, I, r)		Yerdona (L, II, r)	Kurubathur (S, III, s)	
	Progressive development in developing areas						
	Various development programmes in under developed areas			Bidarakere (M, II, m)			Naravi (L, III, m)

The letters in the bracket show:
　　Village size in population;　L: Large, M: Medium, S: Small.
　　Hierarchy by centrality;　I: Lowest, II: Lower, III: Key settlement.
　　Degree of recent changes;　r: Rapid, m: Moderate, s: Slow.

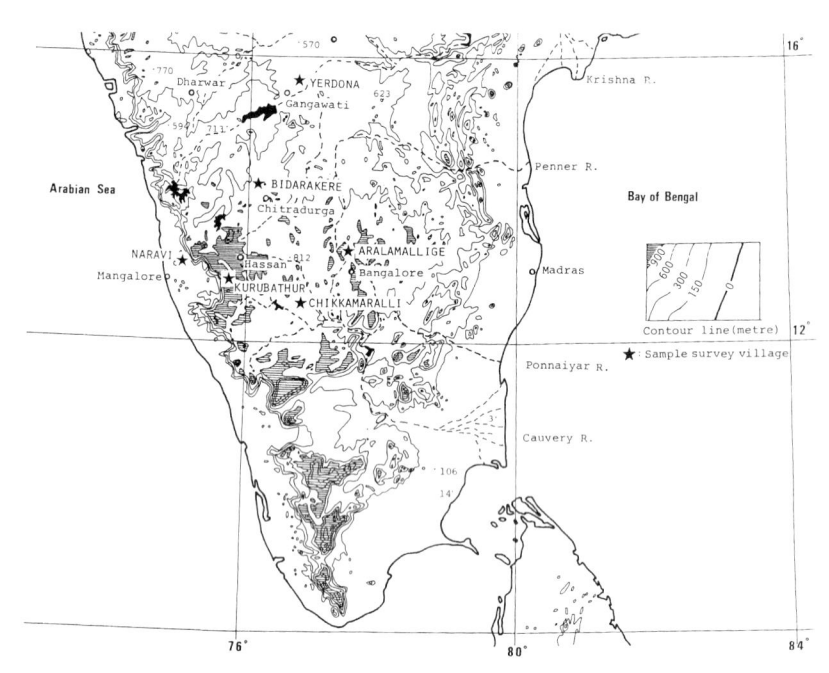

The sample survey villages in 1978, 1980 and 1982

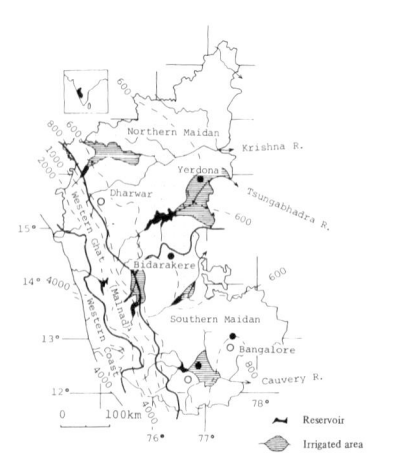

Sketch map of Karnataka state

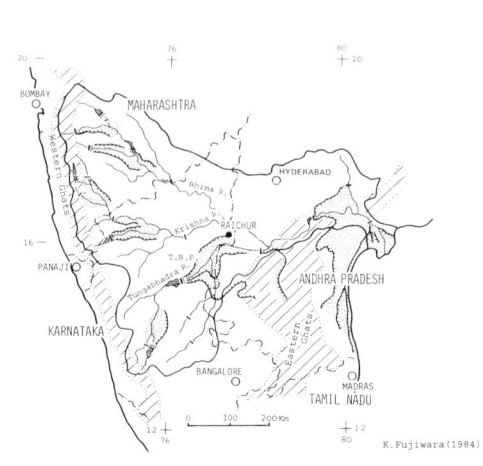

Irrigated areas by canal in the Krishna river basin

phenomenal development in agricultural sector up to the present times. Unless, otherwise we could not have seen such an ever-expanding Indian economy in recent years as we see.

In that context agriculture has been an achilles and, at the same time, playing a role of cruicial role in boosting Indian economy. Agriculture have laid the foundations of recent economic prosperity in India, paving the way to new era for further development.

January 10, 2010
University of London
at Russell square

Iwao Maida

References

Balasubramayan, K. (ed.)(1966): *Village survey monograph – Yerdona village, Census of India 1961, Mysore XI*. Delhi, 96p.

Boudeville, J.R.(1966) *Problems of regional economic planning*. Edinburgh: Edinburgh University Press. Perroux, F. 1955: Note sur la notion de pôle de croissance. *Économie Appliquée* 7: 307–20.

Epstein, S. (1973): *Yesterday, today and tomorrow, Mysore village revisited*. London, 273p.

Fujiwara, K. (ed.) (1980): *Geographical field research in South India, 1978*. University of Hiroshima, 233p.

Fujiwara, K. (1981): Agricultural and rural developments in India. *Chiri (Geography)*, vol. 26, no. 7, pp. 13–23.

Fujiwara, K. (1982): Geographical Field Research in South India, 1980. 178p. University of Hiroshima.

Fujiwara, K. (1984): Geographical Field Research in South India, 1982. 273p. University of Hiroshima.

Fukutake, T. (ed.) (1962): *Social structure in rural India*. Institute of Developing Studies, Tokyo, 238p.

Maida, I. (1981): Structural changes in the socio-economic aspects of villages in South India. *Chiri (Geography)*, vol. 26, no. 7, pp. 44–57.

Ohji, T. (1981): Millet farming in South Deccan Plateau. *Memoirs of the Faculty of Letters*, Kyoto University, no. 20, pp. 1–90.

Singh, R. L. (ed.)(1972): *Rural settlements in Monsoon Asia*. National Geographical Society of India, Varanasi, 510p.

Yonekura, J. (ed.)(1973): *Recent changes in settlements in India*. Kokon-shoin, Tokyo, 505p.

CHAPTER 1
RECENT TRENDS IN THE SOCIO-ECONOMIC STRUCTURE OF VILLAGES IN THE CENTRAL KARNATAKA

Introduction

India can well be described as an agricultural country par excellence. At the same time, it has been often pointed out that rural India has been stricken by continuous vicious circle of poverty, mulnutrition and impaired labour power. The progress of Indian economy towards self-sustained growth has been rather limited and the economic system inherited from the British ruled regime has remained practically unchanged. After the independence in 1947 Indian economy has still remained predominantly agrarian with 73 percent of labour forces depending mainly on land yielding meagre harvests. So many people in rural India have been locked up in the vicious circle mechanism and forced to eke out a living by doing overtiring work of exigous nature, whipping up an exhausted body with hunger.

All out efforts have been made to produce some sizable yields from agricultural land on both sides government bodies and farmers, but for the most part, shabby livelihood seems to have remained substantially unchanged.

Agriculture is absorbing substantial chunk of additional increase in labour forces. They can hardly contribute to the increment of marginal productivity, labour supply in agricultural sector is redundant, and disguisedly unemployed with marginal productivity close to zero.

Under the ever-increasing population pressure on land, combined effects of oversupply of labour and land fragmentation come into play. The land produces diminishing returns for additional labourer who works it. Development policy of rural India ever taken was beset by the overriding difficulty which is generally termed as "population explosion".

In order to level up socio-economic well-being, population in rural area has to be kept in check by appropriate measures just before growth rate comes up to a ruinous level, unless otherwise all the additional net gains from land will be almost certainly wiped out.

In addition to the socio-economic aspects of poverties-ridden rural India, unevenly distributed rainfall mainly caused by wayward coming of monsoon creates serious problems. Rain that comes in fits and starts from year to year. It fluctuates varying from 20 to 40 percent on average, in a lean year crop failures can be seen so common, thus followed by farmine and diseases close behind.

In these agro-climatic setting, wayward pattern of rainfall especially in the drought prone area causes many a disastrous calamity to villagers in the drier area, rate of evapotranspiration tends to rise to the extreme resulting in salinization of the soil and a turbulent, swift-flowing torential rain leaches out valuable minerals and nutrient of the soil.

This, combined with the fact poorly managed contour-bunding system which is prepared for soil conservation makes vulnerable soil in the semi arid zone infertile. Consequently, farmer, dependent on the rainfed agriculture is destined to get all adrift at the mercy of extremely variable and unreliable precepitation.

In this context agricultural sector is considered to be achilles heel of Indian society par excellence, playing an integral part of vital importance in nourishing national economy.

The primary importance of agriculture among the various sort of industries grossly outweight its structural vulnerability, this point must be rightly taken into consideration, in short, level down of performance of agricultural sector leads enevitably up to the serious paralysis of socio-economic system on an expanding scale.

Now that agricultural development has come out to be the mainstay of national concerns and critical importance, five years plan has been spelled out successively to get out of vicious circle of poverty, laying special emphasis on averting back and forth see-saw effects which are derived from virulent whirligig, going around the axis of economic growth and population.

In the case that rate of population growth is going up well over economic growth rate as clearly observed in India, the utterly abject poor shall be reduced to a more deplorable circumstances. This phenomenon might be

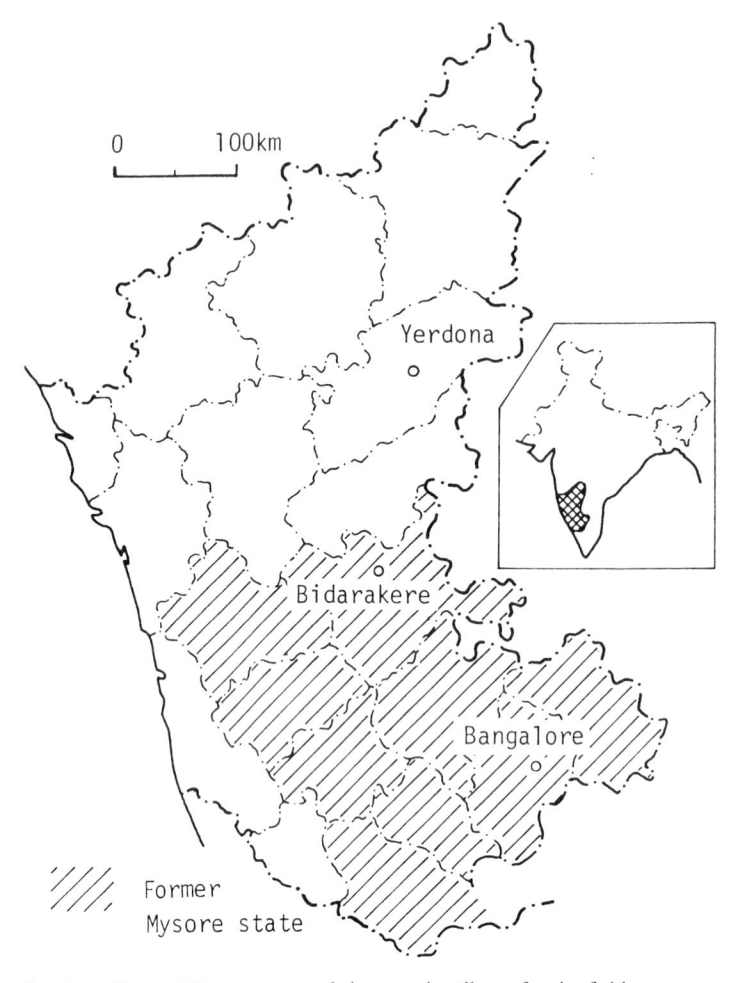

Fig. 1–1 Former Mysore state and the sample villages for the field survey

a potential peril. The main aim of policy makers under the influence of Indira Gandhi regime lies in the near future to come that effective and nuts-and balts counter measures against vicious circle of poverty must be taken to ignite dynamic development for everyone.

For the very purpose, they have renewed their commitment to strengthen and vitalize agricultural activities, notorious whirling of poverty has to be shattered and further more needed to be replaced by an expanding sustained economy. The problem of hunger and hunger and poverty are very severe and deeply routed in rural India.

In this treatise, the author has paid much attention to the multi-facetted agricultural problems with some of workable hypothesis, applied during the field survey, which might be helpful to bring a dynamic structure into sharp focus. The author has traced up what has really been going on in the agricultural activities in Karnataka state in South India, and came to recognize some identifiable trends have cropped out in the recent years.

For that purpose, Bidarakere, the village depending on dry farming was singled out for the field survey which give us precise picture of traditional rural settlement and farming practices. Another village, Yerdona was selected in order to get a close-up view of the newly developed village showing astounding rapid progress with the introduction of irrigation canal system. Profit-oriented farmers are going to maximize their income from land on the competitive base and deeply involved in the market (Fig. 1–1).

Once two villages eked out livelihood from dry land in the semi-arid zone but what makes Yerdona get abreast of Bidarakere in the economic development can be identified as the irrigation system igniting incredible growth potential by which socio-economy is expected to be fully fledged.

This lesson teaches us how the irrigation system has been so helpful as a break-through against the trap of an stagnant economy in the drier area. All the details of comparative analysis of regional characteristics and socio-economic aspects of two villages shall be described as follows.

1. Population in Bidarakere

1.1 Growth of Population

In the first place, it is necessary to compare identifiable trends in population growth in Bidarakere with that of district, state and national levels respectively. Table 1–1 gives the details of long term trends in population increased on different levels. Annual rate of population growth from 1951 to 1981 in India is estimated to be 2.14%, Karnataka state is 2.19%, Chitradurga district is 2.41% respectively. By contrast with these figures, annual growth rate in Bidarakere is estimated to be only 1.04%. Calculating the same rate over 80 years from 1901 to 1981, the results are 1.31% for state, 1.54% for district. Bidarakere is much lower rate 0.86%.

Absolute size of population has been expanding gradually with exceptional period from 1911–1921 population decreases from 1,388 to 1,365, and sex ratio goes up to 104.3. This had been caused by the various epidemic diseases as influenza prevailing widely over the state. As male had been susceptible to the epidemic diseases, relative ratio male to female goes up. Diasynchronic analysis reveals that annual growth rate of population in Bidarakere stands in marked contrast when compared with these of district, state, and nation. Bidarakere has witnessed slow rate of population growth. The stagnant trends in population growth in Bidarakere validate partly low carrying capacity of land.

According to the Trewartha's classification which is in itself a modified form of Köppen's schema, greater part of Interior Karnataka falls into the dry climatic zone (BS), especially is identified as tropical steppe or semi-arid zone. The area embraces a long corridor of semi-arid tropical belt situated in the south of the tropic of Cancer and eastern side of the Western Ghats and the Cardamon Hills. This rain-shadow area, known as a farming zone of India has long been vexed with highly unreliable and scanty rainfall. Water deficit amounts well over to 100cm.

Bidarakere is situated in the drought-prone area (Fig. 1–2). Annual precipitation calculating from 50 years' average is estimated at 638mm in Chitradurga and rainfall fluctuates extremely year by year mainly caused by wayward monsoon with 30–40% deviation from the normal average value. Consequently economic activities is mainly dependent on the traditional rain-fed dry farming. Population size is thus confined within the low level of agricultural productivity.

1.2 Structure of Population

Population data is based on the intensive field survey, which was carried out in 1980. The schedule has been applied to all the households on the door to door method. Population size in Bidarakere adds up to 2,577 which is divided into 1,277 males and 1,300 females. Bidarakere is a village of considerably large size. Numbers of household is 445, out of which 412 households are engaged in agricultural activities. Person per household is 5.79 and number of *jati* amounts to 23.

The table of population structure classified by *jati*, sex and age shows the fact that there can been seen marked difference of shares among *jati*s (Table 1–2). The most predominant *jati* is *Lingayat* with a share of 35%, followed by *Naika* and *Harijana* having a same shares of 13% level. *Adi-Dravida* comes forthly, *Reddy* and Muslim accounts for about 6%. Taking socio-economic aspect of each *jati*s into account, it comes to clear that *Lingayat, Naika, Reddy* and Muslim have played an influential roles tabulated by the size of agricultural holding.

Table 1–3 indicates that predominant *jati* group as *Lingayat, Naika* and *Reddy* are concentrated on the large classes over 20 acres. Whereas one forth of household has only 2.5 to 4.9 acres and nearly one forth is classified as a landless household. Other noticeable point on the *jati*-wise household composition is the fact that scheduled caste group occupies comparatively a substantial share. *Harijana* comes a top 60 household with the 343 population, 13.4% followed by *Adi-Dravida, Adi-Karnataka* with a share of 8.2%, 0.6% respectively, adding up to 23% as a whole. This group is considered to be so called "Weaker section" from socio-economic view point in the village with meagre land, in some cases, landless.

In summing up salient features of population structure Bidarakere can be identified as large and multi-*jati* village in population size and diversified composition of *jati*.

Lastly age specific sex ratio reveals that rate of female is higher in the age group ranging from 0 to 24 except 15–19 class, going up in the upper middle aged group from 40 to 49. And sex ratio stands high on the average in the age class over 50 on the while ratio of female to male gets closer to parity level 101.8. When compared sex ratio of 0–4 age group with that of 5–9 age group the rate of female itself shows considerable drop. Possible reason for this drop is ascribed to high rate of male mortality and sudden decrease of female rate in the age class from 50–69 can be attributed to short interval between child-bearing periods. Woman can not restore her physical strength because of successive child-bearing. In addition to this, they manage to do so many chores inside a house and pain taking hard works in the field along side of male. Thus combined effects come into play, rendering them shortlived.

From socio-economic point of view, especially agricultural holding size, *jati* tends to polarise into the upper class and marginal or landless cluster. Dividing line between two predominant groups can be clearly seen on the 5 and 20 acres level.

Table 1–1 Population trends

	India			Karnataka state			Chitradurga district		
	Population (x10^6)	R	r	Population (x10^3)	R	r	Population (x10^3)	R	r
1901				13,055			511		
1911				13,525	1.0360	0.003543	564	1.1037	0.009915
1921				13,378	−0.0109	−0.636418	575	1.0195	0.001933
1931				14,633	1.0938	0.009006	657	1.1426	0.134199
1941				16,255	1.1108	0.010563	726	1.1050	0.010034
1951	361.08			19,402	1.1936	0.017849	868	1.1956	0.018025
1961	439.23	1.2164	0.019783	23,587	1.2157	0.019724	1,094	1.2604	0.023413
1971	548.16	1.2480	0.022401	29,299	1.2422	0.021925	1,397	1.2780	0.024829
1981	683.00*	1.2460	0.022749	37,071*	1.2653	0.023810	1,773*	1.2691	0.024117
(1901–81)						0.013131			0.015383
(1951–81)			0.021474			0.021856			0.024093

R : Decade growth rate of population ($\frac{P_n-1}{P_n}$)

r : Annual growth rate of population in each decade.

Notes:
  * The estimated figure by Bureau of Economics and Statistics, Government of Karnataka.
  ** The estimated figure by the intensive field survey as of 1980.
  The figure in bracket refers to the results of preliminary population census, 1981.
  Sources:
  Statistical Abstract of Karnataka 1976–77. Gvt. of Karnataka, Bureau of Economics and Statistics, BES. No.47, Bangalore, India.
  Survey Reports of Selected Villages (Bidarakere). Series 14 – Mysore part-VI c, Census of India, 1971.

Annual growth rate of population has been computed from the following formula.

$$\log P = t \log (1+r) + \log P_0$$

Assuming the growth rate is R_n, and annual growth rate is r_n, this relation is expressed as

$$R_n = \frac{P_n-1}{P_n} - 1$$

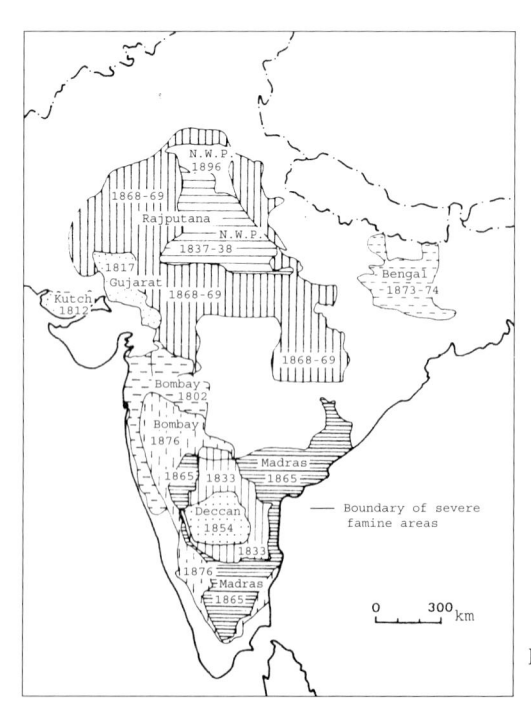

Fig. 1–2 Severe famines of India in 19th Century (From B.M. Bhatia, 1967)

Bidarakere						
Population	R	r	Male	Female	Sex ratio (F/M·100)
1,312			664	648	97.6
1,388	1.0579	0.005647	699	689	98.6
1,365	−0.0167	−0.664158	668	697	104.3
1,542	1.1296	0.012261	791	751	94.7
1,711	1.1096	0.010454	890	821	92.3
1,911	1.1169	0.011117	959	952	99.3
2,158	1.1293	0.012234	1,099	1,059	96.4
2,445	1.1330	0.012565			
2,577**	1.0539	0.005850	1,277	1,300	101.8
(2,748)	1.1193	0.011334			
		0.008582			
		0.010364			

Then, we can get N by solving the following formura.

$$\log(1+r) = \frac{1}{n}\log(1+R_n)$$

Annual increase rate of population in Bidarakere from 1901 to 1980 is to be computed as :

$$\frac{2577}{1312} = 1.964177 - 1 = 0.964177$$

$$\frac{1}{79}\log(1+R79) = \frac{0.29318}{79} = 0.0371114$$

$$1 + r = 1.00858183$$

$$r = 0.00858183$$

The geometric average of annual increase rates of population between two decades (1901-1980) has been computed as follows ;

$$\sqrt[79]{\frac{L_{11}}{L_{01}} \cdot \frac{L_{21}}{L_{11}} \cdot \ldots \cdot \frac{L_{81}}{L_{71}}} = 1.963933$$

$$\sqrt[79]{\frac{L_{81}}{L_{01}}} = \sqrt[79]{1+R_{80}} = 1.00858026$$

1.3 Structural Changes in Population

Population in Bidarakere has been increasing in size from 2,277 in 1962 to 2,577 in 1980. In the consequence, the ratio of infant population (0–14) has dropped from 44.5 to 36.6 and ratio of aged population has slightly decreased from 7.91 to 6.79. Ratio of dependent population has remarkably decreased. Thus substantial burden on the economically active population became less proportionately (Table 1–4).

This fact is clearly reflected in a woman-child ratio, decreasing drastically from 779 to 240. Structural change in population indicates explicitly prominent symptom of a down swing of population pressure on land in the less favored region just as Bidarakere. Fig. 1–3 are depicted from an age specific population structure. The results of an intertemporal analysis can be summarized as follows.

(1) Population pyramid in 1962 is nearly identical with so called pyramid type expanding widely to the downward. This pyramid can be categorized as a "population of high growth potential" type moving on into the late expanding phase.

(2) On the contrary, population pyramid seems to be closer to a pot type which falls into the category of "population of transitional growth". Birth rate is expected to slow down.

(3) Some identifiable demographic transition seems to have occured between 1962 to 1980. One of the most plausible factors is possibly considered to be a change of villagers attitude to a child-bearing, namely, effect of birth control has come into being.

2. Family Structure

In our consideration, the term peasant holding widely seen in rural India denotes an economic production unit of agricultural and stock-raising character, deeply rooted in a local community. And it is also of a size fit to be operated by the labour forces of one family for which it constitutes the main sources of livelihoods.

Agricultural holding is closely related to the notion

Table 1–2 Population structure by *jati*, sex and age class in Bidarakere

Jati	No. of households	0–4 M	0–4 F	5–9 M	5–9 F	10–14 M	10–14 F	15–19 M	15–19 F	20–24 M	20–24 F	25–29 M	25–29 F	30–34 M	30–34 F	35–39 M	35–39 F	40–44 M	40–44 F	45–49 M	45–49 F
Lingayat	144	34	37	54	70	58	71	56	50	40	66	35	34	39	36	28	21	19	20	18	13
Naika	51	13	22	25	25	18	22	14	21	25	26	17	12	10	10	6	10	6	11	4	9
Harijana	60	18	22	25	32	20	26	18	10	22	15	13	17	10	7	13	5	6	7	9	10
Adi Dravida	44	13	13	16	12	14	18	10	8	6	12	12	4	9	7	7	6	8	6	4	5
Reddy	28	8	9	9	4	11	6	12	7	15	15	10	5	5	1	3	5	1	5	3	5
Muslim	26	11	8	15	12	14	10	10	9	7	9	6	6	3	1	3	5	2	3	2	3
Bestan	15	2	2	6	8	3	11	4	4	6	3	2	5	6	4	2	4	–	3	2	3
Madival	14	3	2	2	3	3	1	2	1	3	4	3	2	1	1	1	1	1	3	1	5
Doby		3	2	4	1	2	8	5	3	–	1	1	–	1	–	–	2	1	–	2	3
Vyshaw	6	3	1	1	2	3	–	2	3	3	–	–	1	2	1	–	1	1	–	–	1
Achar	7	1	1	1	1	–	1	1	2	1	1	–	–	1	1	–	1	–	–	2	1
Edigar	4	2	1	2	–	–	–	1	–	3	–	1	2	3	1	1	–	–	–	–	–
Bramin	3	–	1	–	1	2	–	1	4	1	1	–	–	1	–	–	2	–	–	2	–
Golla	2	1	–	2	–	–	3	3	1	2	–	–	–	–	–	1	1	–	–	2	–
Yenna-Setty	2	–	1	1	–	2	3	–	3	1	–	–	–	–	1	1	–	–	1	1	–
Adi Karnataka	2	2	–	–	–	3	–	3	1	1	3	–	–	–	–	–	–	1	–	–	–
Elagar	2	–	1	2	1	1	1	–	1	–	–	1	–	–	–	1	1	–	1	–	–
Adigar	1	–	–	–	–	–	–	–	1	2	1	–	–	–	1	–	–	–	–	1	2
Kamategiru	1	–	–	–	–	2	1	–	–	–	–	–	1	–	–	–	1	–	–	–	–
Vaishava-Setty	1	–	3	–	1	–	–	–	–	–	–	–	–	–	1	–	1	–	–	–	–
Jamagama	3	–	–	–	–	1	–	1	1	–	1	1	–	–	–	1	–	–	–	–	–
Jambini	1	1	1	–	–	–	–	–	–	–	–	1	1	1	–	–	–	–	–	–	–
S T	9	–	–	1	1	1	–	–	–	–	–	–	–	–	–	–	–	–	1	–	–
*	19	1	2	3	4	6	4	5	6	5	3	4	5	1	–	1	2	2	–	3	3
Total	445	116	129	169	178	164	186	148	136	143	161	107	96	93	72	69	67	50	61	57	63
		245		347		350		284		304		203		165		136		111		120	
%		9.51		13.46		13.58		11.02		11.79		7.88		6.40		5.28		4.31		4.66	
Age-specific sex ratio		111.2		105.3		113.4		91.9		112.6		89.7		77.4		97.1		1.22		1.11	

* Unknown

of peasant family, its coherence, internal structure with rhythmical charges in agricultural activities and mentality of its member. These time-honoured structure of an agricultural holding and peasant family has remained unchanged for long times. In this context, analysis of family structure and its identical trends in recent years bring clearly multi-facetted rural problems under consideration into focus.

In an attempt to make clear an internal structure of the 35 sample families in Bidarakere, much attention has been paid on how to identify various kind of family types and to estimate its possible relationships with agricultural landholding size and furtherly to predict a path that certain types of family are most likely to follow in the near future to come. After scrutinizing critically the details of these aspects, we turn our eyes on the relationship between family size and inheritance bearing closely on an agricultural landholding size.

2.1 Family Composition

The internal structure of 35 sample families have been identified by breaking down all the members of each family into 8 categorical kins as a resident group, as seen in Table 1–5. The results classified by the agricultural landholding size has been tabulated in Table 1–6 and illustrated graphically in Fig. 1–4.

high as compared with wife of the head of household (H.H.) and grandson on the average, followed by wife of son and parents with the same proportion 4.2%. The ratio of co-residence with so called co-lateral relatives amounts to 6.3%. The family size tends to grow as landholding size goes up. Average number of person per family seems to be comparatively high.

The family composition classified by agricultural landholding size, with special reference to the five categorical kindred group namely, children (F), grandson, children (F), colateral relatives and parents can be reduced to the generalized pattern showing the relative frequency of each categorical kindred as a co-resident group in Fig. 1–5.

The identifiable trends clearly recognized in the generalized patterns are as as follows:

(1) There are two peaks (I and II) of female chil-

50 - 54		55 - 59		60 - 64		65 - 69		70 - 74		75 - 79		80 - 84		85 - 89		90 - 94		95 - 99				Total	%
M	F	M	F	M	F	M	F	M	F	M	F	M	F	M	F	M	F	M	F	M	F		
15	11	13	8	12	14	5	5	9	5	3	1	4	6	–	–	–	3	–	–	442	471	913	35.43
5	5	6	3	5	4	1	2	4	8	–	1	1	2	–	–	–	–	–	–	160	193	353	13.70
9	2	2	3	6	4	2	4	2	2	2	2	–	–	–	–	–	–	–	–	177	168	345	13.39
2	3	3	1	6	3	1	–	2	–	–	–	–	1	–	–	–	–	–	–	113	99	212	8.23
7	3	1	–	2	–	–	1	–	4	–	–	–	1	–	–	–	–	–	1	87	72	159	6.17
3	4	2	1	3	2	–	1	1	–	1	–	–	–	–	–	–	–	–	–	83	74	157	6.09
2	3	2	1	–	–	–	–	2	1	–	1	–	–	–	–	–	–	–	–	39	53	92	3.57
–	1	2	1	–	1	1	–	–	–	–	–	–	–	–	–	–	–	–	–	23	26	49	1.90
–	–	1	–	1	–	2	–	–	–	–	–	–	–	–	–	–	–	–	–	20	23	43	1.67
1	1	1	–	–	–	1	–	–	–	–	–	–	–	–	–	–	–	–	–	18	11	29	1.13
–	1	1	–	–	–	–	–	–	–	–	–	–	–	1	–	–	–	–	–	9	10	19	0.73
2	–	–	–	–	–	–	–	–	–	–	–	–	–	–	–	–	–	–	–	12	7	19	0.73
–	–	–	1	–	–	–	–	–	1	–	–	–	–	–	–	–	–	–	–	7	11	18	0.70
–	–	–	–	–	–	–	–	–	1	–	–	–	–	–	–	–	–	–	–	11	6	17	0.66
–	–	–	–	–	–	1	1	–	–	–	–	–	–	–	–	–	–	–	–	7	10	17	0.66
–	–	–	–	1	1	–	–	–	–	–	–	–	–	–	–	–	–	–	–	11	5	16	0.62
–	–	–	–	–	–	–	–	–	–	–	–	–	–	–	–	–	–	–	–	6	5	11	0.43
–	–	–	–	–	–	–	–	–	–	–	–	–	–	–	–	–	–	–	–	3	5	8	0.31
–	–	–	–	–	–	–	1	–	–	–	–	–	–	–	–	–	1	–	–	4	3	7	0.27
–	–	–	–	–	–	–	–	–	–	–	–	–	–	–	–	–	–	–	–	1	6	7	0.27
–	1	–	–	–	–	–	–	–	–	–	–	–	–	–	–	–	–	–	–	4	3	7	0.27
–	–	–	–	–	–	–	–	–	–	–	–	–	–	–	–	–	–	–	–	3	2	5	0.19
–	–	–	–	–	–	–	–	–	–	–	–	–	–	–	–	–	–	–	–	2	2	4	0.16
–	2	1	–	–	–	1	–	–	2	–	1	1	–	1	–	–	1	–	–	35	35	70	2.72
46	37	35	19	36	30	15	14	20	24	6	6	6	10	1	–	1	4	1	1				
83		54		66		29		44		12		16		1		5		2		1,277	1,300	2,577	
3.22		2.01		2.56		1.13		1.71		0.47		0.62		–		0.19		0.01					100.00
80.4		54.3		83.3		93.3		120.0		100.0		60.0		–		400.0		100.0		101.8			

dren. The peak of class I clearly shows a growth of a consanguine family on a initial stage. Some considerable drop from class I to class II indicates an increasing number of "married out" daughter to be independent from a family.

(2) As for male children, belonging to the lineal descendents who have equal rights to claim to agricultural property both in kind and cash, show a growing tendency to line with the family, which tends to grow eventually into the extended family or joint family in the due course to be followed.

In the lower classes, they are inclined to hive off from their family. One of the trustworthy reasons is the fact that they are well qualified for claiming to the property of parent (M), but there is a meagre of land to be inherited among the male children, if they want to stay with their consanguious family they can not live just depending only parents means of livelihood. So the things naturally goes that they are expected to be independent from their parents.

(3) The group of grandson grows bigger especially in the class III up to IV. This trend reflects the fact that male children coming of age naturally get married and bears a child as time goes by, and this, combined with a positive feeling in favor of living together with their married or unmarried sibling waiting for coming marriage, because of economizing per head expenses in daily life and patriarchally centered feelings get those family members more closer lineally and colaterally. In this setting, grandson is to be fed, educated and socialized by "family of orientation".

(4) The ratio of living together with parents goes upward slightly in the higher classes. These family is usually identified as jointed or extended family with more than 2 to 3 generations.

(5) The colateral relatives as a brother, sister, nephew, niece and their children of H. H. tends to rise as landholding size goes up. In this category, once married-out sisters of H. H. who came back native home are included. This type of family serves as a family of refugee securing her to make a living in some ways.

2.2 Family Types

Taking all the facts and points so far discussed into

Table 1-3 Distribution of household by *jati* and land holding size in Bidarakere

Jati \ Acre	0	0.1–0.9	1.0–2.4	2.5–4.9	5.0–7.4	7.5–9.9	10.0–14.9	15.0–19.9
Reddy				6	1	3	4	–
Lingayat	20		7	24	26	14	21	10
Nayaka	23		4	14	7	2	7	
Hariyana	23	1	6	11	6	1	2	1
Adi Dravida	2		10	18	9	1	3	1
Muslim	5	1		4	4	4	5	3
Bestar	7		2	1	1	1	2	1
Janagama	1							2
Ediga				1	1	1		1
Bandari Kayakaru						1		1
Yenna-Setty							1	1
S T	2			6			1	
Madival	4			6	2	2		
Golda					2			
Elagar				1	1			
Vysyara	3			3				
Achar	5		1	1				
Adigar				1				
Kamategiru				1				
Bramin	3							
Adi Karnataka	2							
Barikaru	1							
Janbini	1							
Vaishava-Setty	1							
*	1		1	12	1		1	
	72 (33)	105	2	31	112	62	30	47	20
	105 (33)	23.60	0.45	6.97	25.17	13.93	6.74	10.56	4.50

* Unknown
( ) The figures in brackets refer to non-agricultural households.

consideration various kinds of families have been categorized into the several basic types by the number of household and generation in a certain family. After identifying family type, this has been arranged according to the landholding size. Each result has been presented in Table 1–7 and Table 1–8 respectively to predict a possible shift of family types.

In the first place, as for Table 1–7, it is of much interest to note that the family consisting of one generation with two households which is categorized as A–2 has the highest percentage of all types amounting to 14 families, followed by B–3 and then A–3. The other types than these types are very few. Age of H. H. is going to get older as number of generation and household increases. The oldest H. H. appears in the A–3 type, next to it, B–3 and C–4 come at the nearly same age. Age differences among the seven classes suggest what type is most likely to follow alternative paths. Distribution of sample families classified by the family types and agricultural landholding size is shown in Table 1–8. Predominant types in each landholding size are clearly seen. In the class I the predominance A–1 and A–2 types is apparent, and in the II class A–2, A–3 and B–3 type appear to be of same frequence. When it comes to class III, and IV, A–2 and A–3 type are relatively prevalent when going up to the biggest class V, major pattern is going to split into A types and B or C types.

In total, A type is adding up to 23 families, 65.7% followed by B type, 10 families, 28.6%. Only 2 families belong to C type with the weight of 5.7%. However, it should be noted that C type is not always exceptional and thus negligible case. Far from being negligible, this type suggests some reliable classes as will be discussed carefully, with special reference to so called time-honoured Hinduism and inheritance system observed strictly in this village. All the types observed here have to be examined in the context that a phenomenal frequency of the categorized family types would be well explained by the comparative study of developmental processes of family cycle process.

20.0–24.9	25.0–29.9	30.0–34.9	35.0–39.9	40.0–44.9	45.0–49.9	50.0–54.9	55.0–59.9	
6	1	1	1	2	1	–	1	28
6	7	3	2	1	1			144
1		1	1					51
								60
								44
								26
								15
								3
								4
								2
								2
								9
								14
								2
								2
								6
								7
								1
								1
								3
								2
								1
								1
								1
								16
13	8	5	4	3	2	–	1	445
2.92	1.80	1.12	0.89		1.35			

3. Inheritance System and Family Types

This section presents a comparative analysis of the closely connected socio-economic relations between inheritance system and family type vis-a-vis inheritance system. The author has carried out a intensive survey to find out the predominant inheritance customs observed in the village, asking H. H. to single out the nearest one from the following five answers what is the best way of inheritance of your farm land. The result has been summarized as follows:

(1) To give all the land to one successor .... 2(5.7%);
(2) In some way to divide the land among the sons of H. H. .... 30(8.5%);
(3) In some way to divide the land among the children of H. H. .... 1(2.8%);
(4) To leave the matter to the surviving family .... 1(2.8%);
(5) Others or no idea. .... 1(2.8%).

The frequency of inheritance customs to divide the land among the sons of H. H. adds up to 30 households, amounting to 85.8%. The case of (1) falls into the category of promogeniture confining the heir exclusively to the issue especially, the eldest son. There is two H. H. who selected the case (1) of alternatives insisting on the point that they could not divide all the property among male heirs just because he did not have even a single boy. The one has physically handicapped female and the other has only one daughter who is not in principle qualified for the inheritance of H. H.'s property. If they have male children, they are naturally inclined to single out the case (2). In this sense, the share of the case (2) is going up to 91.4%. Now it comes to be clearly apparent that nearly almost H. H. tends to prefer dividing the property of land among his sons.

The inheritance system could be identified as being unilineal basically tracing affiliation of each member to the inheritance through patrilineal linkage. This trend as these patrilineally centered inheritance system can be observed in the facts that co-resident group is ex-

Table 1-4 Age-specific structure of population in Bidarakere

	Population	Age-specific composition 0-14	Age-specific composition 15-59	Age-specific composition >60	Index of age-specific structure *	Index of dependents population Total	Ratio of infant population	Ratio of aged population	Woman-child ratio **
1962	2,277	1,014(44.53)	1,083(47.56)	180(7.91)	17.75	110.25	93.63	17.75	779
1980	2,577	942(36.55)	1,460(56.66)	175(6.79)	18.58	76.51	64.52	18.58	240

* $I_{as} = \sum_{65}^{99} Pa / \sum_{10}^{14} Pi$

** $R_{tf} = \sum_{0}^{4} P / \sum_{15}^{44} Pf \cdot k$ (1,000)

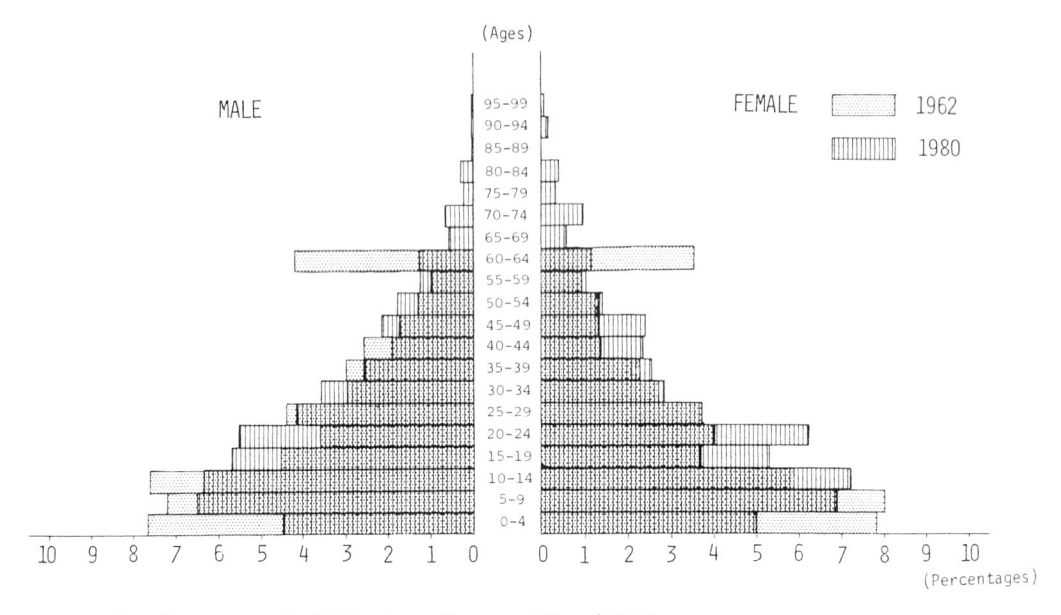

Fig. 1-3 Population pyramid of Bidarakere village in 1962 and 1980

tended bilaterally to the corelatives of H.H. and that the ratio of co-residence of male children with their kindreds is going up as the holding size increases. As far as the details and particulars pertaining to the inheritance and residence obtained from field survey are concerned, a family is well organized by the patriarchial inheritance and residence rules. It is very interesting to note that the socio-economic effects of the patriarchial inheritance system on the family type and landholding size. Intertemporal comparison of three cadastral maps showing subdivision of landholding in Bidarakere in 1902, 1956 and 1980 reveal striking features on how slow but steady population pressure on the land have seriously affected the subdivision of agricultural landholding over nearly eighty years (Fig. 1-6 A, 6B, 6C).

Two types of plot in their shape are clearly shown on the map of 1902. The one is identical of a block shaped lots concentrated in the southeastern and northwestern parts of the village. And the other is a stripshaped fields around the settlement, running along east-west axis.

As the subdivision of agricultural landholding has been accelerated, the extent of lot has become small, and at the same time, the shoe-strip shaped lots have come to be narrower and elongated, and as regarding block fields have come to be closer to a nearly strip shaped field.

Another more characteristic feature is a successive subdivision of the land, situated between the irrigation tank and main road passing by the settlement to the Jagalur. A handful peoples, who have deeply rooted in the village community and played an influential role of vital importance, grow a rice on a meagre land irrigated by the tank. They stake their social prestages and honour on clinging to the paddy field, where much more inputs of labour, capital and incessant careful attentions are needed to cultivate it. When coming down to the microscopic level, it comes to be much more clear how widely those small plots scatter over and the close examination of particular data on land holding gives some reliable informations about how the subdivision of landholding has come into being. The results are summerized in the Tables 1-9 and 1-10 and this, in turn has been depicted on the map (Figs. 1-7A, 7B, 7C). Taking the typical cases, the households, 100 and 140 out of nine agricultural holdings drawn on the cadastral map for instance, fragmented plots of small size field, scattering widely over within and outside the village can be clearly seen in both cases. In the first place, the H. H. of 140 has the land of 35.8 acres in total dividing into 12 plots (Table 1–9).

Within the village he has 24.4 acres consisting of 8 plots, and 11.4 acres, 4 plots in Matadayamav-vanahalli. The average extent of plot in Bidarakere amounts to 3.05 acres and 2.85 acres outside village. On the whole, this figure comes out to be just only 2.98. He has been purchasing eight small plots successivelly from 1951 up to present. The purchased fields amounting to 19.1 acres in total has been addend to the inherited land. This is a few case H. H. has managed to succeed in acquiring additional land. He has nearly doubled agricultural holding in the thirty years. Family type of this holding falls into the category B-3.

As for the case of No.100 (Table 1–10), one of the most well-off landowners has 56 acres in total, field is divided into fifteen plots in 3.75 acres. Corresponding figure in Bidarakere is calculated to be 2.59 acres. He has purchased 29.7 acres since 1965. The ratio of the purchased land to the grand total is 52.8%. This adopted H.H. has strenuously made his all out efforts not only to expand the agricultural holding in size but also maintaining the patrilineally extended big family. He has four male children who are old enough to keep a house for themselves independently. Some of them have been very impatient to get along outside the family. However H. H. has long been earnestly anxious to build up a well-off family by coaxing gently his sons out not to leave the family. He has played an integrating role for setting up a congenial family full of solidarity and confabulation. If some of his sons left family, sooner or later this family would have broken up disjointedly because the fission of a big family of an extended or jointed types would be accelerated by the growing tendency of individualism among the co-heirs which directly lead to the subdivision of land under these inheritance system.

On the contrary to these cases, the case of No.119–1 shows how quickly the agricultural land has been reduced to the minimal extent in size under the same inheritance system over three generations. The H.H. of this household had inherited 8 acres with the same share among his 5 brothers. Female is to be excluded for the inheritance (Fig. 1 8).

In this way, the original land of 48 acres had been divided into just only 8 acres after the death of H.H.'s father. Though he has accumulated small plots from 8 to 19 acre 30 guntes. This land has divided among the 8 heirs with the extent of 1.0 to 2.0 acres. The total amounts to 13.5 acres. The rest of the land has been sold for not only the dowry paying to the sprouse of female children. But the repaying the loan for daily expenditure. As daughters have got married successively, the amount of dowry adds up to 21,000 rupees

The original land of 48 acres has thus been reduced nothing more than the meagre land varying from 1.0 to 2.0 acres over three generations. The more children H.H. has and the less land he has, thus the more poor he would be.

In other words, admitting that the other condition being equal as the size of agricultural landholding which constitutes the bulk of landed interests in a peasant economy depends mainly on the numbers of male heirs, the original holding tends to be divided and subdivided into pieces from generation to generation as long as the inheritance system and the dowry customs are strictly observed.

The compound effects caused by these system and manners come into play apparently as a decisive deterrents checking many a people in this village in the detestable vicious circle of poverty.

The big family as an extended or joint family has long lived out as a negative counteractive device to the rural impoverishment. Big family are very few in number and concentrated in the highest class of agricultural class, however, this implicitly indicates the fact that the H.H. of big family had come to consider one of the main reasons of poverty to be nothing else but subdivision of land under the growing population pressure on the land. This is a major reason why they organized tactifully themselves against it. Patrilocal rule has thus keenly observed in a higher class in Bidarakere.

We have to pay much attention to the advantageous aspects of subdivision and fragmentation of land as well as harmful consequences as an increased cost of production, underutilization of capital and labor inputs and disputes and litigation caused by fencing, making pathway involving law courts.

Creation of economic equity through equal distribution of land could be counted one of the advantageous points. Under primogeniture or ultimogeniture system poralization of the have and the have-not tends assumably occur. A coexistence of massive class of landless laborers and a handful of big landholders in an extreme case should be a drawback to an egaritarian community. Subdivision has led to a widespread distribution of land property, achieving economic equality to some extent and degrees.

Further more important points are the fact that subdivision and dispersal of land equalize not only differences of soil types and fertility in values, but acts as an insurance against the total failure of crops due to the capricious monsoon. In fact, the equal rights of male heirs claiming to the property have helped to create egaritarian community and acted as a social insurance in the double sense. The one is insurance, against crop failure by equal-sharing of different types of soil and fertility, and the security lies in the name of independent family and community itself. Family and community

Table 1–5 Family structure and family types classified by a categorical kinship as a co-resident group

Household No.	1	2	3	4	5	6	7	8	9	10	11	12	13	14	15	16	17
One-person household		1															
H.H.	1		1	1	1	1	1	1	1	1	1	1	1	1	1	1	1
Wife of H.H.	1		1	1	1	1	1	1	1	1	1	1	1		1		
Eldest son	1		1	1	1	1		1	1	1	1	1		1	1	1	1
2nd son				1	1	1		1	1			1	1				
3rd son						1		1	1			1					
4th son								1									
5th son																	
6th son																	
7th son																	
8th son																	
Female	1		5	2	4	2		3	2			3	1				4
Wife of son								1	1	1		1		1	1	1	
Grandson								1	2			1	4	4	7	4	
Parents																	
Grand parents																	
Brother and sister					1	1									1		
Nephew and niece															1		
Others																	
Non-relatives Co-resident servant																	
No. of family member	4	1	8	6	9	8	2	10	9	6	4	10	7	7	11	9	6
No. of household	1	1	1	1	1	1	1	2	2	2	1	2	1	1	2	1	1
No. of generation	2	1	2	2	2	2	1	2	3	3	2	3	3	3	3	3	2
Age of H.H.	26	48	49	45	40	40	27	50	58	80	28	58	74	65	65	75	50
Family type	A-2	A-1	A-2	A-2	A-2	A-2	A-1	B-2	B-3	B-3	A-2	B-3	A-3	A-3	B-3	A-3	A-2
Landholding size	1-30	2	2	2-30	3-2	4	4	4-20	5	6	7	7	8-20	8	11	12	14

are well organized against rainy days to extend helpful hand to the weaker section in a stagnant subsistence economy.

As the cross-sectional analysis of family types in Bidarakere seems to suggest some identifiable trends in the developmental phases of the family type classified by the number of household and generation. The developmental phases could be divided into the three periods in which cyclical shifts from expansion to replacement of families by way of fission. Some of characterizations and salient features of family in each periods from view point of developmental phases of family could be reduced to the following.

(1) Expansion period: Conjugal family consisting of newly married couple gives birth to children. In this period, nuclear family grows into a big patriarchally extended or joint family by degrees over time.

(2) Fission period: Some of male children remain to stay with their parents and unmarried sibling, after coming of age they get married and create an additive nuclear family within an extended or joint family. The rest of male children leave home for establishing another households. The house head is going to marry a marriageable daughter off to a man of same caste with handsome dowry. Additive fission of family occurs in this way. However, in the case of an extended on joint family, the members of co-resident group are so nicely amalgamated into a steadfast socio-economic unit by the patriarchal leadership that they are less motivated to hive off. This holds true of especially male heirs.

By contrast, some family belonging to lesser class in holding size, most of H.H. can hardly direct their personal existence, and still less rules over peacefully all the wills of a family member. They tend to be supportive of a voice of patriarchally extended or joint family head and seem attune cordially to it.

(3) Replacement: Generally speaking, the fission of household comes first in a patriarchally extended or joint family. The household of this family type remains, unchanged for more than 3 generations as long as he is left alive, and works robustly as hard as ever. Even if he is dead, someone of male co-heirs is thought to manage to get along with them, and tends to keep agricultural holding as an integrated set of going concerns without division of property in close cooperation with his brothers. They seldom if ever, divide the inherited property among them. But, in reality, it has been keenly felt that there are some discrepancies among co-

18	19	20	21	22	23	24	25	26	27	28	29	30	31	32	33	34	35	Total	
																		1	
1	1	1	1	1	1	1	1	1	1	1	1	1	1	1	1	1	1	34	35
1	1	1	1	1	1	1	1	1	1			1	1	1	1	1	1	30	30
1	1	1	1	1	1	1	1	1	1		1	1	1	1	1	1	1	31	
1	1	1	1	1	1	1	1	1	1			1	1	1	1	1	1	23	
1	1	1	1	1	1		1	1				1	1	1	1			16	
1	1	1	1	1			1					1		1	1			10	88
	1		1									1			1			4	
	1		1									1						3	
	1																	1	
1	2	2	2		3	3	1	2	4			2	2	1		2	4	58	58
			1			1			1					2				12	12
						2		1		2				6				34	34
	1		2					1	1		1	2	2	1	1			12	12
			1				2		4						1			11	
							3		1									5	18
			1											1				2	
7	12	8	14	6	9	7	9	13	10	7	6	7	12	9	15	11	8	287	
1	1	1	3	1	2	1	2	1	1	1	2	1	2	2	3	1	1	49	
2	3	2	3	2	3	2	3	2	3	2	3	3	3	3	4	3	2	87	
46	63	50	50	40	80	50	50	50	60	24	54	38	48	40	58	46	55	1,780	
A-2	A-3	A-2	C-3	A-2	B-3	A-2	B-3	A-2	A-3	A-2	B-3	A-3	B-3	B-3	C-4	A-3	A-2		
19	23	27	30	30	30	31	35-32	35	35	39-37	40	46	47	45	56-08	61-24	63		

heirs on how to manage household and land properly as well. The problem extends widely over how to get their children socialized and educated which seems to be caused mainly by communication gaps among the household of the younger brother and that of the aged and a different ways of their owns. Various incongruity in these respects is implicitly represented in everyday life, eventually leading to the prolonged division of land. This holds true particularly when they are going to prepare to be independent. No.100 household as has discussed is now on the way to this case. No.100 could be identified as a transitional type of this kind. The sift of family type onward remains to be seen. By contrast, in the case of another type of extended or jointed family of A-2, A-3, B-2 and B-3 replacement of family tends to take place in parallel with a fission of household accompanied by a division of property under the growing influences of individualism among co-heir. The absence of powerful patriarchal leaderships accelerates further replacement and fission processes of a family.

The accurate prediction of possible shifts of family types in the light of recent developments in Bidarakere could make clear inherent factors at work. Fig. 1-9 indicates the simplified schema, showing a predicted sift of family types with average age of H.H. in blanks.

The most possible way to follow, is shown with arrows in black. Taking all the discussion as described above and the average age of H.H. into account, the path from A-1 to A-2 leading to B-2 type and A-3 has been predicted. This shift is followed by the paths from B-2 to B-3 and from A-3 to B-3 finally it comes to C-3 and then C-4 types in the end. Those shifts are accompanied by an increase in the average age of H.H. On the reverse course back to A-1 type this is termed conjugal family (nuclear family without children), C-type is mostly like to go back to B or A type. B, A-2 and A-3 types are also come back to A-1 type. All those possible paths are illustrated graphically on the plane with four quadrants classified by the class averages of family size in number and farming size (agricultural holding including the leased-in land) in the village.

Much attention has been paid to the population pressure expressed in family size in number especially number of male co-heirs who are qualified to the division of property, bearing on the agricultural landholding size.

As clearly seen in Fig. 1-10 there is a triangle section consisting of marginal farmers class and landless ag-

Table 1–6 Family composition classified by agricultural landholding size

Member of resident group	I 0 – 4.9	II 5.0 – 9.9	III 10.0 – 19.9	IV 20.0 – 39.9	V > 40	Total
H.H.	1.00 (16.7)	1.00 (14.0)	1.00 (12.1)	1.00 (10.5)	1.00 (10.3)	1.00 (12.2)
Wife of H.H.	0.88 (14.6)	0.83 (11.6)	0.50 (6.1)	0.90 (9.5)	1.00 (10.3)	0.85 (10.4)
Children	3.75 (62.5)	2.68 (37.2)	3.00 (36.3)	5.40 (56.8)	4.86 (50.0)	4.17 (50.9)
Male	1.63 (27.1)	1.66 (23.2)	1.75 (21.2)	3.50 (36.8)	3.28 (33.8)	2.52 (30.7)
Female	2.13 (35.4)	1.00 (14.0)	1.25 (15.1)	1.90 (20.0)	1.57 (16.2)	1.66 (20.2)
Wife of son	0.12 (2.0)	0.67 (9.3)	0.50 (6.1)	0.20 (2.1)	0.43 (4.4)	0.34 (4.2)
Grandson	–	2.00 (27.9)	2.75 (33.3)	0.30 (3.2)	1.15 (11.8)	0.97 (11.8)
Parents	–	–	–	0.50 (5.3)	1.00 (10.3)	0.34 (4.2)
Colateral relatives	0.24 (4.0)	–	0.50 (6.1)	1.20 (12.6)	0.28 (2.9)	0.52 (6.3)
Total	6.00 (100.0)	7.17 (100.0)	8.25 (100.0)	9.50 (100.0)	9.71 (100.0)	8.20 (100.0)

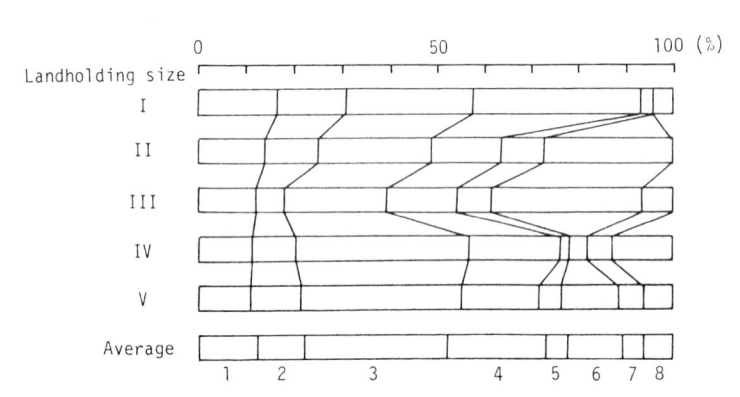

Fig. 1–4 Family composition classified by agricultural landholding size in Bidarakere
1: Household, 2: Wife of H.H., 3: Children (M),
4: Children (F), 5: Wife of son, 6: Grand son,
7: Parent, 8: Colineal relatives

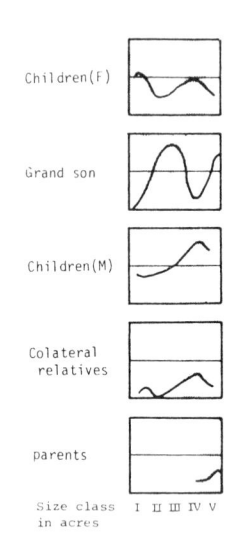

Fig. 1–5 Simplified patterns of the frequency of the categorical kindreds as a co-resident group

Table 1–7 Distribution of the family types and the age of householder classified by numbers of household and generation

Nos. of household		Numbers of generation				χ
		1	2	3	4	
A	1	2 (37.5)	14 (42.3)	7 (60.1)		23 (47.3)
B	2		1 (50.0)	9 (59.2)		10 (58.3)
C	3			1 (50.0)	1 (58.0)	2 (54.0)
x̄		2 (37.5)	15 (42.9)	17 (59.1)	1 (58.0)	35 (50.9)

The ages of H.H. are shown in blanks.

Table 1–8 Distribution of sample families classified by family type and agricultural landholding size

Family type	I 1 – 4.9	II 5 – 9.9	III 10 – 19.9	IV 20 – 39.9	V > 40	Total
A–1	2 ⎫7					2 ⎫
A–2	5 ⎭	1 ⎫3	2 ⎫3	5 ⎫7	1 ⎫3	14 ⎬23
A–3		2 ⎭	1 ⎭	2 ⎭	2 ⎭	7 ⎭
B–2	1					1 ⎫10
B–3		3	1	2 ⎫3	3 ⎭	9 ⎭
C–3				1 ⎭	4 ⎫	1 ⎫2
C–4					1 ⎭	1 ⎭
Total	8	6	4	10	7	35

ricultural labourers class alongside of Y axis. H.H. of these class originally belonging to the higher class has reduced to the most depressed states under the equal division inheritance system. Growing size of family in number specially co-heir tends to aggravate the impoverished state in an intensified way. In short, landholding size is inversely proportional to the number of children, especially male co-heirs under the equal division inheritance system. Population pressure plays a decisive role par excellence in both levelling up and down an economic well-being, which is directly related with agricultural holding size.

Chap.1 RECENT TRENDS IN THE SOCIO-ECONOMIC STRUCTURE OF VILLAGES IN THE CENTRAL KARNATAKA

Fig. 1–6 Sub-division of Agricultural landholdings in 1902, 1956 and 1980

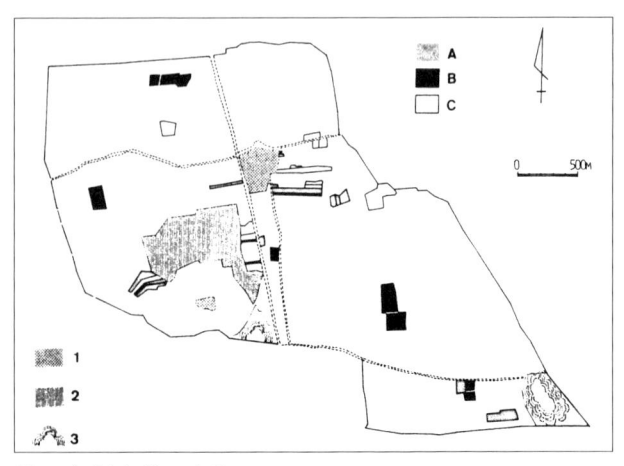

Household A, B and C

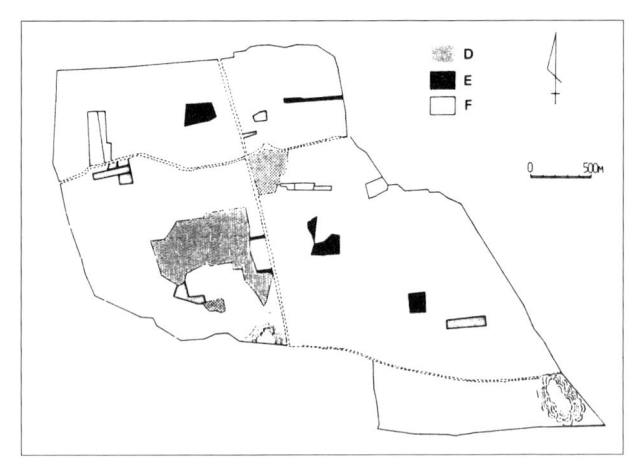

Household D, E and F

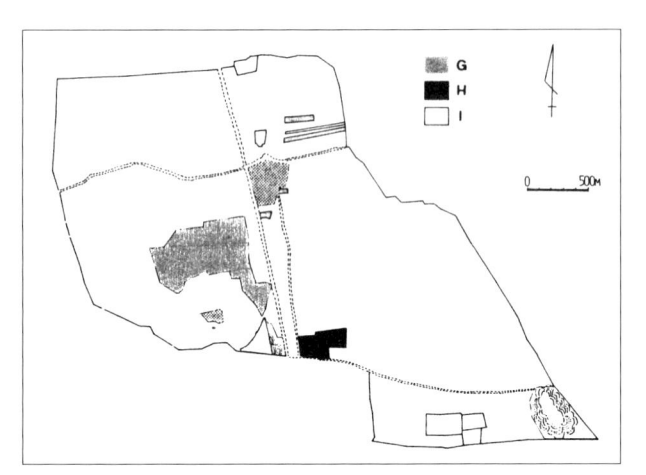

Household G, H and I

Fig. 1–7 A, 7B, 7C
Cadastral maps showing distribution of fragmented plots of the nine agricultural holdings in Bidarakere (1980)
A–1 shows each lands holded by each farmers.
  1: Settlement,
  2: Irrigation tank,
  3: Hillock.

Table 1–9 Distribution of agricultural land (Household No. 140)

Village	Loc. No.	Acre	Major crops			Year of purchasing of land		
						Year	Acre	Price (RS)
Bidarakere	1	11–3BP	0–35	Pa	Jo		1980	0–35	χ
	2	14–6P	0–25	Pa	Jo		1978	0–25	χ
	3	14–11	1–20	Ra	Pa	Be			
	4	74–1P	6–28	Ra	Jo				
	5	94–1	4–16	Ra	Jo				
	6	142–1	4–09	Co	On				
	7	161–1	4–04	Co	On		1966–67	4–04	7,000
	8	163–2P	1–39	Co	On		1959–60	1–39	1,200
Matadadya Mavvanahalli	9	45–2	2–02	Ra	Jo		1965–66	2–02	1,200
	10	48–1	3–18	Ra	Jo		1951–52	3–18	600
	11	48–2	3–21	Ra	Jo		1961–62	3–21	2,100
	12	49–2	2–15	Ra	Jo		1961–62	2–15	725
Inherited land			16–28						
Purchased land								19–04	
Grand total			35–32						

Quadrant III consisting of patriarchally extended or jointed family C–3, C–4 (including a part of A and B types) could be identified as the highest class in agricultural holding enjoying agreeable rural life than any other class. It is interesting to predict the possibilities that C–3, C–4 types should follow the supposed paths. After getting to the turning point A, they are going to downward. And then the one is heading for A the another, while the other turns its course down to B point on the left side. A–A path is explained by the facts that the effect of some agricultural innovations come into play in a compounded way and then enhance productivity per additional inputs. The augumented yields result in the increment of income which is ploughed back to

Table 1–10 Distribution of agricultural land (Household No. 100)

Village		Loc. No.	Acre	Major crops			Year of purchasing of land		
Bidarakere	1	14–1	0–22	Pa	Jo				
	2	14–8	0–28	Pa	Jo				
	3	15–2	2–11	Ra	Jo				
	4	45–5a	2–20	Ra	Jo				
	5	70–26	2–26	Ra	Jo				
	6	72–2	3–38	Ra	Jo				
	7	100–2b	0–03	Ra	Jo				
	8	102–1	0–07	Ra	Jo				
	9	107–1	3–10	Ra	Jo				
	10	144–2	6–31	Co	On	Ch			
	11	143–2	4–00	Co	On	Ch	1,979	(4–00)	4,000
	12	129–2	4–10	Co	On	Ch			
Nibagur	13	37–1	8–13	On	Co	Ch	1,967	(3–06) (5–07)	2,000 5,000
	14	36–1	6–35	On	Co	Ch	1,965	(6–35)	5,000
Rastemakunte	15	34–1P	10–20	Ra	Jo		1,943	(10–20)	400
Inherited land			26–20						
Purchased land								29–28	16,400
Grand total			56–08						

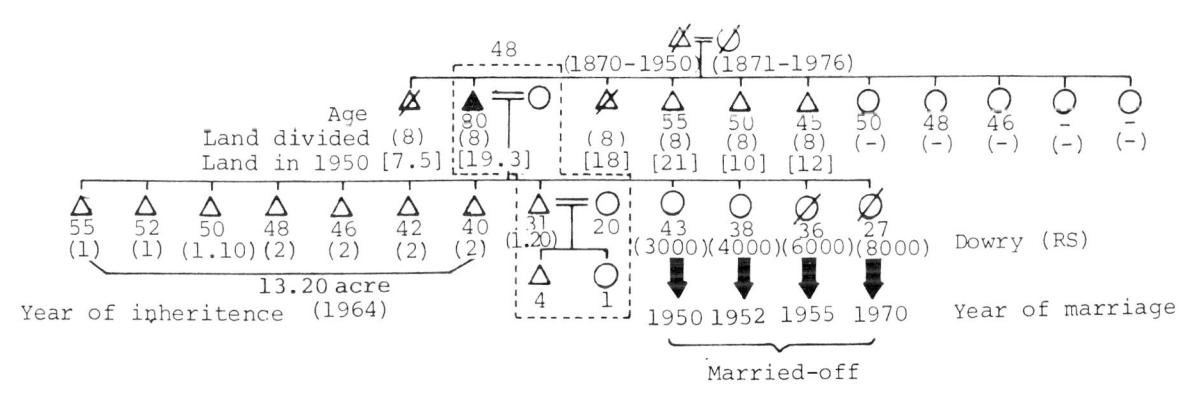

Fig. 1–8 Typical example showing subsequent process of subdivision of agricultural landholding in Bidarakere

purchase of additional land. This process is accompanied by the decrease of dependents. The unwanted child is expected to be reduced to nil by the birth control.

The A–B path gives more realistic picture onward, permitting there will be no identifiable innovation emphasizing agricultural performances with the fertility rate remain unchanged. To make the thing more worse, growing sentiment of individualism favorably received among the co-heirs is most likely to divert its course to the left. This results in hiving off of co-heirs and thus a fission of family. The same can be said on the down swing to the point C. In this way, following recurring cycle is going on just like a pinwheel.

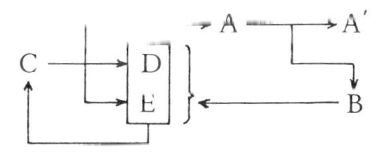

Even if holdings get smaller, some of the costs incurred by agricultural activities such as expenses of maintaining his family, a pair of bullocks and a few of agricultural implements do not decrease proportionately. Dwindling holding size thus results in driving bulk of people into a tight corner. They can hardly afford even to live from hand to mouth, and still less feed another mouth. They call down curses profanely on an invisible enemy causing hateful poverty with and inquiring frown. In this context, the envisaged scene would be in the deadlock, far from being promising. There is no specific cure giving an immediate effect for the betterment of their socio-economic conditions.

4. Recent Trends in Agricultural Management and Its Performance

The main theme of this section is to clarify the recent trends in cropping patterns, agricultural management and its economic performance revealed in the sample holdings with a reference to an access and availability of water needed for agricultural activities.

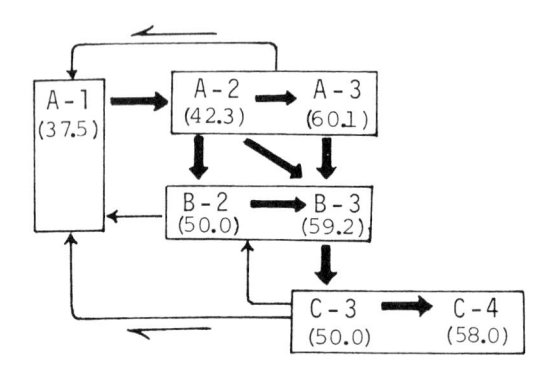

Fig. 1–9 Simplified schema showing a predicted shift of family types
Blanks indicate the age of householder

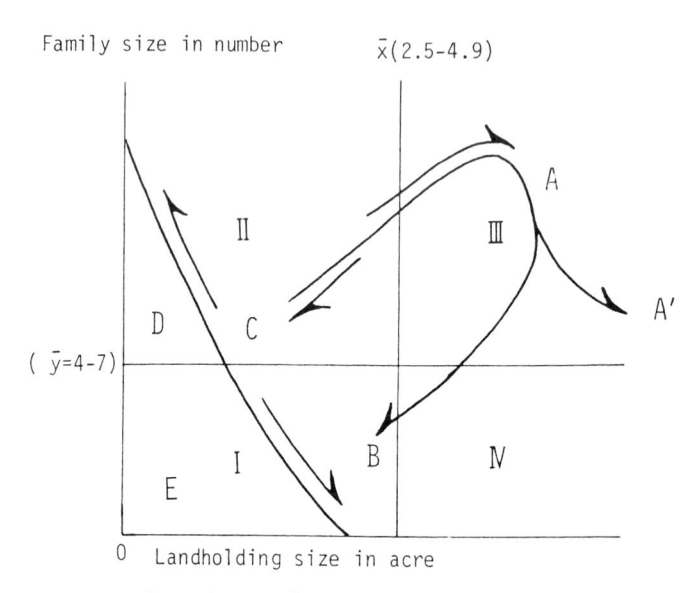

Fig. 1–10 Relation between family size and landholding size and its relevance to predicted shifts of family types

4.1 Agro-climatic conditions ; an overview

The National Commission on Agriculture of India has studied *taluk*-wise rainfall data of Karnataka state and identified 26 rainfall patterns, which has been reduced to the four agro-climatic regions, viz, Coastal belt, West Ghats region, Transitional belt and Dry belt. The study area falls into the category of Dry belt, indicating rainfall patterns of E_4 (C_3 D_1) D_1 E_3 in Raichur and D_1 E_3 (C_1 D_3)D_1 E_3 in Ganga-vathi respectively (Fig. 1–11). The *taluk*-wise patterns of annual rainfall illustrated in Fig. 1–12 indicate that annual rainfall totals to 521.6mm in Gangavathi and 544.9mm in Jagalur respectively.

Amount of rainfall of Jagalur *taluk* is little more than that of Gangavathi both in pre-monsoon and post-monsoon periods. Rainfall more than 50% of annual rainfall is concentrated in monsoon period from June to September. Insufficient and unevenly distributed annual rainfall has long affected seriously the bulk of poverty stricken farmers who could not but depend mainly on the rain-fed dry farming seen just as that of Bidarak-ere in Jagalur *taluk*.

Vagarious onset of monsoon aggravate further growing conditions of various crops. Cultivators in a drier belt put it as follows, if *aridra* fails, *daridra* befalls. "*Aridra*" rain, usually starting from fourth week of June down to first week of July, is supposed to prognosticate fair weather for growing crops. *Daridra* literally means poverty. For cultivators, it is of vital importance to for see a possible onset of *Aridra* and a timing of cultivation, to which their livelihood is mainly left.

A time-honored proverb as this saying has long believed to be one of wisdoms to the poor who could barely live out by adapting themselves to severe natural environment. By contrast, in the Yerdona where agro-climatic conditions are similarly severe as in Bidarakere, a greater part of farmers have been benefited in every ways on an easy access to water supply for agricultural and daily use by canal irrigation system.

Agro-climatic data and agricultural production of the study area shows that the area under irrigation is esti-

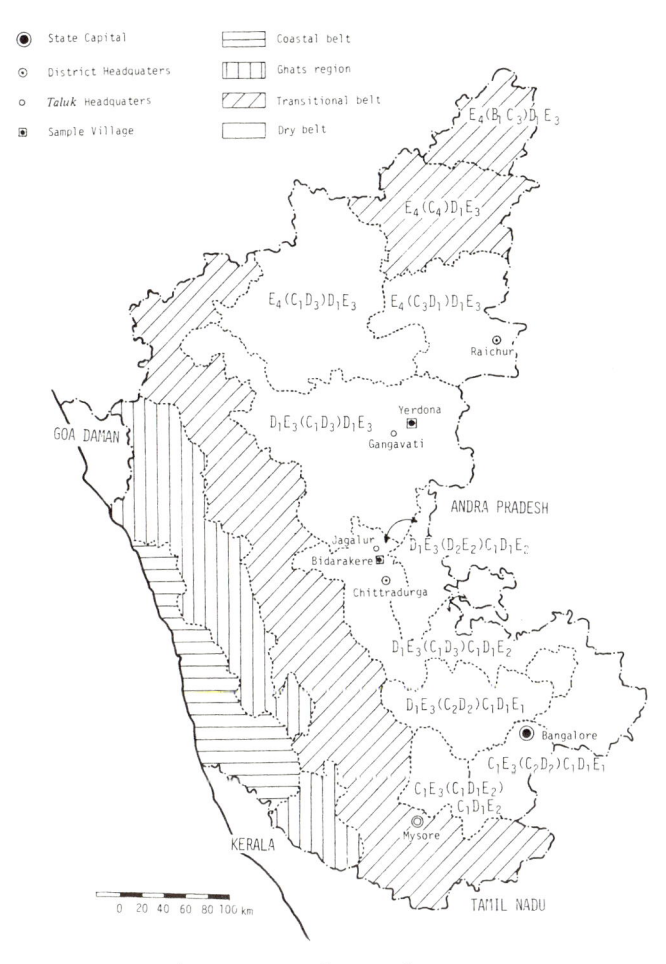

Fig. 1-11 Agro-climatic region of Karnataka state

Source: Cropping pattern for different rainfall patterns and agro-climatic regions of Mysore state, 1973, The University of Agricultural Sciences, UAS Tec. Series, No.1.

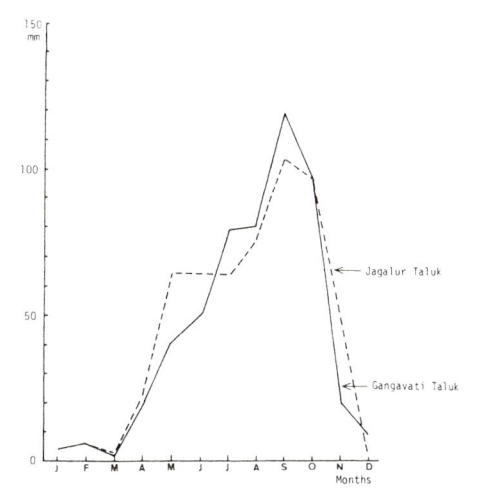

Fig. 1-12 Rainfall in Jagalur and Gangavathi

The yearly rainfall is given for the 3 seasons as follows:
Rainfall;
    Pre-monsoon (2–5) – Described in alphabets and numbers preceding the brackets
    Monsoon (6–9) – Described in alphabets and numbers within the brackets
    Post-monsoon (10–1) – Described in alphabets and numbers succeeding the brackets
    Example
$$\underbrace{D\,E}_{\text{Pre-monsoon}}\underbrace{(C\,D)}_{\text{Monsoon}}\underbrace{D\,E}_{\text{Post-monsoon}}$$

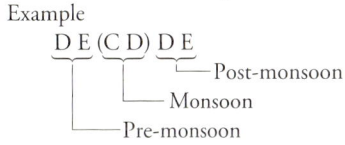

Coding for quantity of rainfall;
    A; 30 cm/per month, B; 30–20, C; 20–10, D; 10–5, E; Less than 5.
Coding for number of month ;
    Number (affix of alphabets) 1 to 4 refers to the number of month receiving the specified rainfall.

mated at 36% in Gangavathi and only 8% in Jagalur respectively. Soil type of each *taluks* is identified as medium black soil in Gangavathi, and deep black soil in Jagalur. As for cropping pattern, paddy ranks well a head of other crops, followed by the commercial crops as cotton and groundnuts in Gangavathi.

By contrast, small millets, *jowar, ragi* mainly for home consumption stand out in various crops the fields in Jagalur *taluk* stands thick with these drought-resident crops and productivity of these crops is comparatively high. Marked contrast in the cropping patterns between two *taluks* could be attributable to availability of water (Table 1–11).

4.2 Cropping Patterns in Bidarakere

The vacuity of the open and gently undulating land extends over the farthest bounds of the horizon here in Bidarakere the extremely parched land grows a variety of crops well over twenty mainly based on mixed cropping system (Fig. 1–13). A variety of crops could be reduced to the three major crops, viz, corns mainly for home consumption as *ragi, jowar,* and pulses of drought resident and high yielding varieties, commercial crops as onion, chili, cotton and highly profit-oriented crop just as mulberry for sericulture and coconuts.

The first group crops are grown on the northern part of the settlement which is mainly composed of red soil. Commercial crop in second and third groups are grown on a more fertile black soil, which extends over southern part of the village with the gradual development of high-yielding varieties and a wide variety of drought resistent strains in the last two decades, they came widely into use among the farmers. However these varieties absorb much of fertility from relatively infertile land.

They began to realize the vital importance of fertility-augmenting crop combination. There has been widely held notion that fertility-absorbing crops must be followed by fertility-maintaining or even augmenting crops just as cowpea and other pulses. They have tried to find out in their own way a sequence of crop combination to avoid squeezing a few drops of marginal yields

Table 1-11 Agro-climatic data and agricultural production of the study areas

		Raichur District				Chitradurga District			
		Gangavati *Taluk*				Jagalur *Taluk*			
	Month	Rainfall				Rainfall		Rainy days	
Pre-Monsoon	J	–	63.5 (12.2%)			4.1	98.2 (18.0%)	0.3	6.7 (17.2%)
	F	–				6.6		0.3	
	M	1.3				3.1		0.3	
	A	21.3				22.9		1.7	
	M	40.9				65.5		4.1	
Monsoon	J	51.5	334.0 (64.0%)			48.3	291.8 (53.6%)	4.1	23.2 (59.6%)
	J	80.9				63.7		7.0	
	A	81.8				75.9		5.9	
	S	119.8				103.9		6.2	
Post-Monsoon	O	95.6	124.1 (23.8%)			97.3	155.0 (28.4%)	5.5	9.0 (23.2%)
	N	20.4				45.5		2.7	
	D	8.1				8.1		0.5	
	Total	521.6	(100.0)			544.9	(100.0)	38.6	
Area under irrigation		36%				8%			
Soil type		Medium black				Deep black			
	Crop	Area covered in ha	Yield (kg/ha)	Index I *	Index II **	Area covered in ha	Area covered	Index I *	Index II **
Agricultural Production	Paddy	22,000	2,400	237	193	–			
	K. *Jowar*	14,000	740	75		9,000	1,300	184	257
	R. *Jowar*	3,000	–	145	84	3,000			
	Bajra	4,000	–	101		2,000			
	Small millets	9,000	–			12,000			
	Tur	1,000	–	61		3,000			
	Other pulses	9,000	–			5,000	370		144
	Sugarcane	3,000	–			–			
	Groundnuts	15,000	460	93	157	2,000	560	96	87
	Cotton	18,000	490	59	121		390	58	127
	Other oil seeds	1,000	–			2,000			
	Ragi					7,000	1,010	111	91

* Productivity Index to national average
** Productivity Percentage to state average

out of meagre land, unless otherwise, fertility of land would have been exhausted completely, but the most appropriate crop combination remains to be seen at present.

Meanwhile, commercial crop as onion, chili and cotton are supposed to have played an important role in levelling up a gross income of agricultural holdings in some ways. These crops are admittedly the most profitable one. But at the same time price of these crops tend to fluctuate extremely, thus resulting in an instability of agricultural incomes.

The more farmer gets involved in market, the more he shall be all adrift. In this sense, cultivation of these crops might be a kind of speculative business. Consequently the appropriate crop combination depends on a pay-off balance among soil fertility of soil, marketability, profitability of crop and stability of agricultural incomes.

Another commercial crops are as mulberry and coconuts have come into being recently. These crops have just been exclusively cultivated by a handful affluent farmers of the upper class. They are still staggering in its infancy. Because vast amount of investment in construction of irrigation well, pump and piping facilities, and it will thus take much time to depreciate those fixed costs as well as running costs of capital assets. In addition high level techniques involved both in cultivation and management are greatly needed for a successful performance of agricultural holding as a going concerns. It will take so much time for business of this kind to evolve into an full fledged level.

4.3 Comparative Analysis of Crop-combination Types in Bidarakere and Yerdona

A wide variety of crops are grown extensively on a dry land by mixed cropping method. Mixed cropping is one of the most important farming practices, sowing main crops with one or two subsidiary crops on the same land. Mixed crops are sown by seed drills the mixed crop as *ragi, jowar*, cotton and onion are usually grown with three lines of the main crop and one line of the subsidiary crop or five lines by one line.

Bewilderlingly complicated cropping patterns in numbers and varieties as clearly seen in Fig. 1–13 could be orderly reduced into the simplified form, viz, crop-combination type being reduced by paying attention to the regular occurrence of prevalent type and similarity of internal connectivities in it. Results are shown in

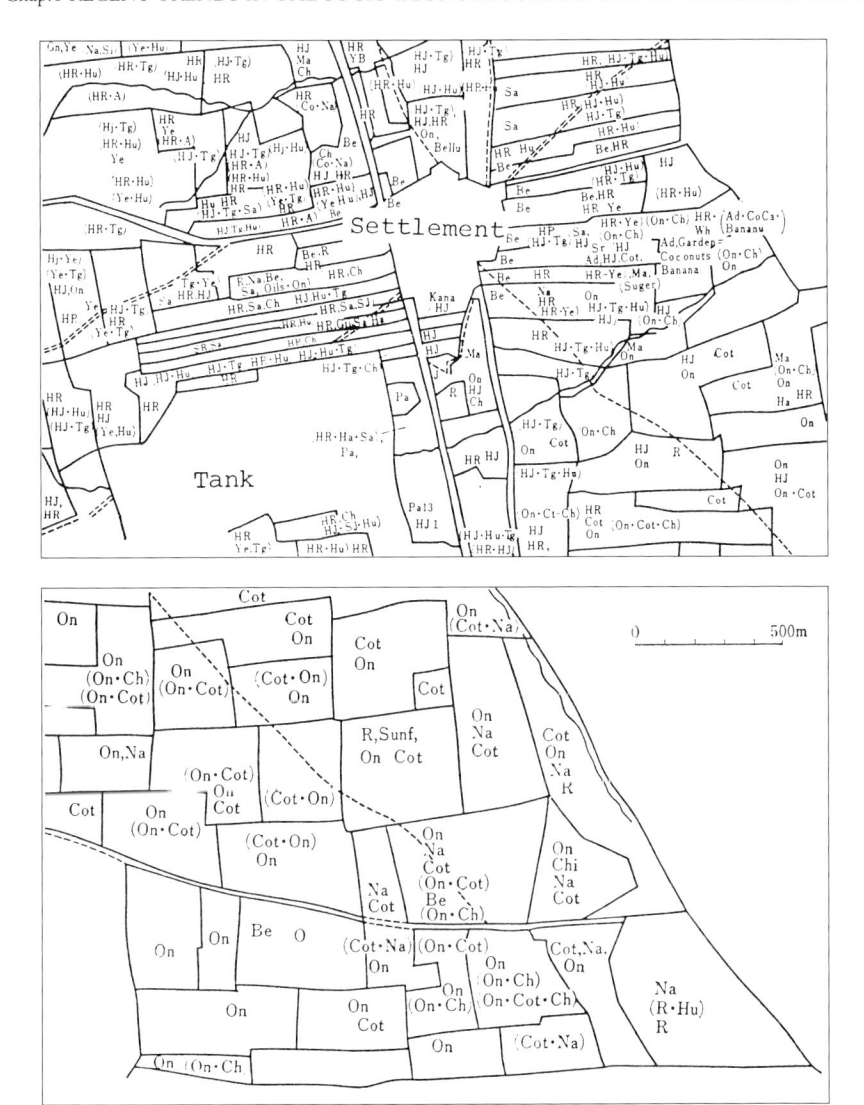

Fig. 1-13 Mixed cropping patterns in Bidarakere (1980)

R: *Ragi*, HR: Hybrid *Ragi*, J: *Jowar*, HJ: Hybrid *Jowar*,
Pa: Paddy, Sj: *Sajje*, Ma: Maize, Na: *Navane*, Wh: Wheat,
Tg: *Thogari*, Hu: *Huluri*, A: *Avare*, Gn: Groundnut,
Gu: *Gralu*, On: Onion, Ye: *Yellu*, Ne: *Neragire*, Ca: Castard,
Sf: Sunflower, Ad: *Adeke*, Cot: Cotton, Sr: Sericulture,
Ch: Chili, Be: Barren land.

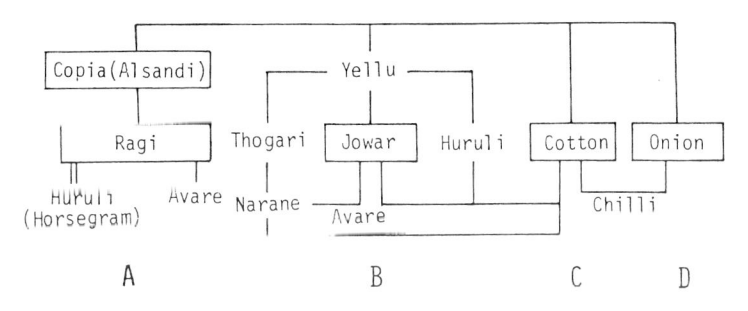

Fig. 1-14 Major crop-combination types in Bidarakere

Fig. 1-14.
Mixed cropping patterns in Bidarakere are shown in Table 1-12. Dominant crop mixtures could be divided into four groups in which major crop is combined with minor crops in an unified way that profitability is compatible with maintaining fertility of land.

The reason why a fertility of soil in this village has not long been exhausted could be partly due to mixed

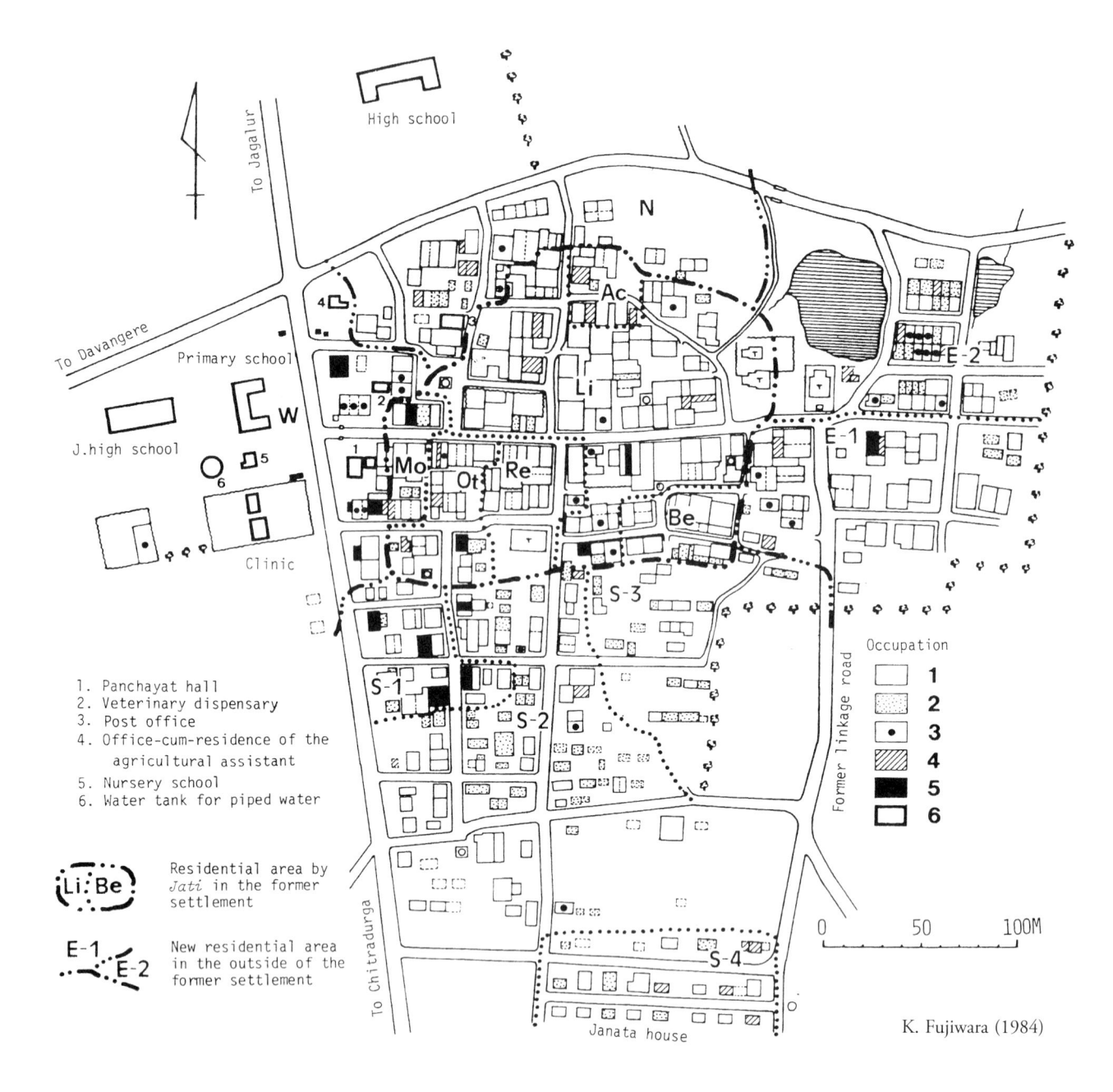

Fig. 1–15 Settlement pattern in Bidarakere

Residential area by *jati*
Li: *Lingayat,* Re: *Reddy,* Be: *Bedda*
Ot: Others, Mo: Moslem.
W, N, E-1, etc. Shows a block of new residential area respectively.

Occupation
1: Agriculture
2: Agricultural labour, coolie and marginal farmer (below 2 acres)
3: Non-agricultural workers (Teacher, Driver, tec.)
4: Artisan
5: Shop, hotel
6: Public facilities

Chap.1 RECENT TRENDS IN THE SOCIO-ECONOMIC STRUCTURE OF VILLAGES IN THE CENTRAL KARNATAKA

Tungabahdra Dam ; 49m H, Length 2,450m, command area covers some 500,000ha

Settlement landscape Bidarakere, situated drought pronc area

Table 1–12 Crop-combination types in Bidarakere and Yerdona

		Bidarakere		
		Crop-combination type	Land holding size	Area cultivated
1	B–3–108	Y(1–3/4)	1.30	1.30
2	B–2–29–2	$\begin{bmatrix} R \\ Hg \end{bmatrix}(2)$	2.00	2.00
3	B–3–122–2	R (1)		
4	B–3–84–1	$\begin{bmatrix} C \\ On \\ Ch \end{bmatrix}(2.30)$	2.30	2.30
5	B–3–54	R (3–1/2)	3.20	3.20
6	B–3–66	$\begin{bmatrix} R \\ Hg \end{bmatrix}(4)$	4.00	4.00
7	B–3–67	$\begin{bmatrix} C \\ On \\ Ch \end{bmatrix}(4)$	4.00	4.00
8	B–3–70	R (4.20)	4.20 (2.00)*	4.20
9	B–3–10	R (3) – G (2)	5.00	5.00
10	B–3–59	R (4) – J (1)	7.00 (2.00)*	5.00
11	B–1–119–1	R (3) – J (2) – $\begin{bmatrix} Co \\ On \\ Ch \end{bmatrix}(1)$	7.00 (3.00)*	6.00
12	B–3–9	$\begin{bmatrix} R \\ Hg \end{bmatrix}(2) - \begin{bmatrix} J \\ Rg \end{bmatrix}(2) - \begin{bmatrix} Co \\ On \end{bmatrix}(2) -$ Na (2)	8.00	8.00
13	B–3–61	Ye (4) – J (3) – R (1–1/2)	8.20	8.20
14	B–3–45–1	R (4) – J (2) – G (2) – $\begin{bmatrix} Co \\ On \end{bmatrix}(4)$	12.00 (5.00)*	12.00
15	B–1–94	R (8.0) – J (3)	11.00	11.00
16	B–2–58	J (6.0) – R (5.0)	63.00 (26.00)**	11.00
17	B–3–113	$\begin{bmatrix} R \\ Hg \end{bmatrix}(4.0) - \begin{bmatrix} J \\ Rg \end{bmatrix}(4.0) -$ Gg (3.0) – Na (1.0)	12.00	12.00
18	B–3–1	Co (7.0) – Ye (3.0) – $\begin{bmatrix} R \\ Hg \end{bmatrix}(3.0) - \begin{bmatrix} J \\ Rg \end{bmatrix}(1.0)$	14.00	14.00
19	B–1–102	$\begin{bmatrix} R \\ Av \\ Ni \end{bmatrix}(4.0) - \begin{bmatrix} Co \\ On \\ Ch \end{bmatrix}(4.0) - \begin{bmatrix} J \\ Hg \\ Rg \end{bmatrix}(4.0) -$ R (4.0) – Na (1/2)	16.20	16.20
20	B–1–17	$\begin{bmatrix} R \\ Hg \\ Rg \end{bmatrix}(10) - \begin{bmatrix} Co \\ On \\ Ch \end{bmatrix}(10) -$ Bj (0.10)	19.00 (5.00)*	20.25

Table 1–12 (Continued)

		Bidarakere	Land holding size	Area cultivated
		Crop-combination type		
21	B–1–63	$\begin{bmatrix} J \\ Hg\ (10) \\ Rg \end{bmatrix} - \begin{bmatrix} R \\ (8) \\ Ni \end{bmatrix} - \begin{bmatrix} On \\ (2) \\ Ch \end{bmatrix} - Co\ (2)$	23.00	22.00
22	B–1–101	$R\ (16) - \begin{bmatrix} Co \\ On\ (8) \\ Ch \end{bmatrix} - Rg\ (4) - Hg\ (2)$	27.00	30.00
23	B–1–110	$R\ (10) - J\ (10) - On\ (5) - Ch\ (3) - F\ (2)$	30.00 (3.00)*	30.00
24	B–1–109	$R\ (12) - \begin{bmatrix} Co\ (4) \\ On\ (3) \\ Ch\ (2) \end{bmatrix} - J\ (5) - Hg\ (2) - Rg\ (1)$	30.00	29.00
25	B–1–114	$R\ (10) - J\ (9) - \begin{bmatrix} Co \\ On\ (8) \\ Ch \end{bmatrix} - Hg\ (4) - Rg\ (1/2)$	30.00	31.50
26	B–2–37	$\begin{bmatrix} Co \\ On\ (12) \\ Ch \end{bmatrix} R\ (10) - J\ (10)$	31.00 (3.00)*	32.00
27	B–2–140	$R\ (10) - \begin{bmatrix} Co \\ On \end{bmatrix}(10) - \begin{bmatrix} J \\ Rg \end{bmatrix}(9) - P\ (2) - Bt\ (1/3)$	35.32	31.33
28	B–2–59	$R\ (10) - Co\ (8) - J\ (6) \begin{bmatrix} On \\ Ch \end{bmatrix}(4) - M\ (2) - F\ (5)$	35.00	35.00
29	B–1–49	$Co\ (6) - R\ (4) - Rg\ (4) - On\ (4) - J\ (3) - Hg\ (1)$ $- Ch\ (1/2) - F\ (6\ 1/2) - Cn$	35.00	29.00
30	B–2–67–3	$R\ (20) - \begin{bmatrix} J \\ Hg\ (10) \\ Rg \end{bmatrix} - \begin{bmatrix} Co \\ On\ (8) \\ Ch \end{bmatrix} - Ma\ (2)$	39.37	40.00
31	B–1–109	$\begin{bmatrix} J \\ Hg \end{bmatrix}(12) - R\ (8) - \begin{bmatrix} Co \\ Ku \end{bmatrix}(6) - M\ (1/2) - Jw\ (1-1/2)$	40.00	28.00
32	B–2–47	$\begin{bmatrix} Ch \\ On \end{bmatrix}(16) - Co\ (12) - R\ (6) - Na\ (5) - Ye\ (5) - J\ (2) - Rg\ (2)$	46.00	48.00
33	B–2–24	$J\ (20) - R\ (10) - \begin{bmatrix} Va \\ Hg \end{bmatrix}(10) - \begin{bmatrix} Co \\ On\ (6) \\ Ch \end{bmatrix}$	47.00	46.00
34	B–2–19	$\begin{bmatrix} Co \\ On\ (15) \\ Ch \end{bmatrix} - \begin{bmatrix} Jo \\ Hg\ (10) \\ Rg \end{bmatrix} - R\ (10) - Va\ (4)$	45.00	39.00
35	B–2–100	$\begin{bmatrix} Co \\ On\ (25) \\ Ch \end{bmatrix} - \begin{bmatrix} Jo \\ Hg\ (13) \\ Rg \end{bmatrix} - R\ (13) - Va\ (4)$	54.38	55.00

Table 1–12 (Continued)

		Yerdona	1	2	3	4
		Crop-combination type	Land holding size	Total area cropped	Area cultivated	Intensity of land use (2/3 × 100)
1	B–5–132	P (0.30)	0.30	0.30	0.30	100
2	B–1–62	J (1.0) = G (1.0)	1.00	2.00	1.00	200
3	B–4–9	P (3.15) = G (3.15)	3.15 (1.20)*	6.30	3.15	214.3
4	B–2–80	J (2.0) – G (1.0) – Na (1/2)	3.20	3.20	3.20	100
5	B–5–17	P (4.0)	4.00	4.00	4.00	100
6	B–3–104	Co (4.0) – G (1.0) = G (1.0)	5.00	6.00	5.00	120
7	B–1–63	$\begin{bmatrix} J\,(1.20) \\ J\,(4.0) \end{bmatrix}$ = G (1.20) – Co (1.20)	7.00	8.20	7.00	121.4
8	B–5–157	[P (1.0) = P (1.0)] – B (0.10)	7.35 (6.25)**	2.10	1.10	190.1
9	B–2–42	Na (8.0)	8.00	8.00	8.00	100
10	B–2–69	J (6.0) – Co (4.0)	10.00	10.00	10.00	100
11	B–3–96	P (5.0) – J (3.0) – S (2.0)	10.00	10.00	10.00	100
12	B–5–140	Bj (9.0) – Bj (5.0)	14.00	14.00	14.00	100
13	B–1–48	$\begin{bmatrix} P \\ P \end{bmatrix}$ (12.0) – G (2.0)	14.00	14.00	14.00	100
14	B–3–41	[Bj (8.0) = G (8.0)] – Na (6.0) – J (2.0)	16.00	24.00	16.00	150
15	B–5–100	$\begin{bmatrix} P\,(6.0) \\ P\,(6.0) \end{bmatrix}$ = $\begin{bmatrix} G\,(6.0) \\ G\,(6.0) \end{bmatrix}$ – Co (4.0)	16.00	28.00	16.00	175
16	B–5–144	Hg (7.0) – J (4.0) – Co (3.0)	18.00	14.00	14.00	100
17	B–3–72	Co (10.0) – $\begin{bmatrix} J\,(2.0) \\ J\,(3.26) \end{bmatrix}$ – G (2.36)	18.22	18.22	18.22	100
18	B–3–102	Co (10.0) G (4.20) – G (3.00) – Bg (1.0) – P (0.20)	19.00	19.00	19.00	100
19	B–3–49	G (18.20) – P (5.0)	23.20 (8.20)*	23.20	23.20	100
20	B–3–78	Co (12.0) – G (4.25) – J (4.0) – P (3.0) – Bg (1/2)	24.25	24.25	24.25	100
21	B–1–94	$\begin{bmatrix} G\,(8.0) \\ P\,(4.0) \end{bmatrix}$ = G (12.0) – J (8.0) – Co (6.0)	26.00	36.00	26.00	146.2

Table 1–12 (Continued)

		Yerdona	1	2	3	4
		Crop-combination type	Land holding size	Total area cropped	Area cultivated	Intensity of land use (2/3 × 100)
22	B–1–8	J (10.0) – Na (10.0) – [P (4.0) – G (4.0)]	32.00	28.00	24.00	116.7
23	B–1–6	P (10.20) – J (10.0) – G (8.0) – Co (7.0)	35.20	35.20	35.20	100
24	B–3–99	$G(18.0) = \begin{bmatrix} G(8.0) \\ Ma(6.0) \end{bmatrix}$ – [P (6.0) = P (6.0)] – J (6.0) – Co (6.0)	36.00	56.00	36.00	155.6
25	B–1–79	$\begin{bmatrix} P(34.0) \\ G(6.0) \end{bmatrix} = \begin{bmatrix} P(6.0) \\ G(30.0) \end{bmatrix}$ – Na (9.0)	46.00	85.00	46.00	184.8
26	B–3–30–1	$\begin{bmatrix} P(20.0) \\ G(20.0) \end{bmatrix} = \begin{bmatrix} P(4.0) \\ G(36.0) \end{bmatrix}$ – J (15.0) – Co (5.0)	60.00	100.00	60.00	166.7
27	B–1–22	$J(20.0) - \begin{bmatrix} Co(6.0) \\ Co(10.0) \end{bmatrix}$ – [J (10.0) = G (10.0)] – $\begin{bmatrix} G(6.0) \\ P(4.0) \end{bmatrix}$ – [Na (4.0) = G (4.0)] – Ga (3.00)	65.28	77.00	60.00	128.3
	Total		524.875	658.925	501.05	131.5

* Leased-in land ** Leased-out land

Abbreviation :

A	*Avare* (field beans)	Be	*Betlvine*	Bg	Bengal gram	B	*Bajira*	Ch	Chili	Cn	Coconuts
Co	Cotton	F	Fallow	Gg	Green gram	Hg	Horse gram	J	*Jowar*	Jw	White *jowar*
M	Mulberry	Ma	Maize	Na	*Navane*	Ni	*Niger*	On	Onion	P	Rice
R	*Ragi*	S	Sugarcane	Rg	Red gram	V	*Varagu*	Ye	*Yellu*		

Explanatory notes :

As for Bidarakere

Crop-combination type represented in the bracket refers to mixed cropping combination.

As for Yerdona

A set of crop-combination in the square bracket as in $[\begin{smallmatrix} A & C \\ B & D \end{smallmatrix}]$, indicating multiple cropping combination, are rotated in a crop year.

Crops A and B refer to first crop. C and D come after in sequence as the second crops.

Crops rotation more than 2 sets of brackets are identified as the symbol (=) showing rotation connectivity of crop groups between and or among brackets.

The figures in blanks refer to the area cropped.

* Leased-in land   ** Leased out land

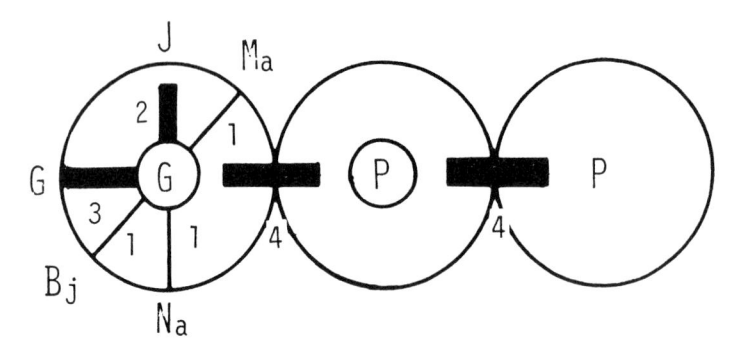

Fig. 1–16 Connectivity among major crop-combinations of the sample holding in Yerdona

cropping maintaining and even more augmenting leguminous component. Leguminous plants such as horse gram (*huruli*), red gram (*thogari*), bengal gram (*kadale*) and field gram have been long played a part of vital importance in maintaining soil fertility up to the present time and saving inputs of artificial fertilizer. Recently much attention has come to be paid cow pea, which is locally called as *copia, alsandi* in Kannada. All those leguminous crops are very useful not only for maintaining soil fertility, but securing foodstuffs for livestocks. Beans that *alsandi* bears are edible.

In addition to those points, mixed cropping system is considered to be a guard against total crop failure due to unfavorable agro-climatic conditions, crop mixture could diversify overall risks to some extent.

The farmers in Bidarakere have adjusted themselves not only to the hospitable land, but actively tried to bring the desolated land with a meagre soil fertility into an tamed domain under their control. We can draw many a lessons from their adaptive processes to the severe natural environment which seems to have been unconquerable at first sight.

To say the least of it, it follows thus that one of most reliable and first hand experiences through paintaking labours has been gradually distilled into a full fledged farming system, viz, mixed cropping practices.

We can read how farmer has maintained ecological balance between man and nature and worked up a fragmentary patch of information how to use a land into a feasible farming system. In trying to cipher out exigent crunch , whereas there has been ever-expanding discrepancy between growing population and staggering productivities of the marginal land. The existing mixed cropping is nothing else but a time honored farmer's wisdom.

In this sense, the mixed cropping stands out first against severe environment for agricultural production.

On the other hand, rural landscape in Yerdona is completely different from that of Bidarakere. Village life has been greatly improved by the canal irrigation.

Agricultural production has taken a very decided turn for the better. It was not 1957 until irrigation canal has been extended to this village.

Availability of irrigated water has led to the cultivation of commercial crops as rice, groundnuts, sugarcane, cotton and others and then put the spurs to use agricultural land in so diversified and intensified ways that farmer could realize farm incomes as high as possible. This paved the way to the betterment of village life. Rural scenes has dramatically changed and taken an amazingly brisk and animated turn for the better.

Agricultural production had once depended on a capricious monsoon rain. The greater part of farm lands are now well irrigated by two distributaries directly connected with Tungabhadra Left Bank Canal. Rain-fed try farming has been drastically replaced by irrigation farming. The wet land is classified into three categories by the crop grown on it and thus required amount of water firstly heavy crop as rice comes, followed by light crops as *jowar*, cotton and ground nuts follows lastly garden called a *baghayat*.

Those crops are supposed to be cultivated under the designated localisation scheme. Wet land which has been brought under irrigation is calculated at 3401.17 acres, 58.5% total cultivable land, 5812.09 acres. Rice as a heavy crop takes leads first, amounting to 1614.15 acres with a proportion of 47.4% to the total wet land.

Prevalent crop-combination type are arranged by land holding size in the right side sheet of Table 1–12. As clearly seen, basic crop-combination could be identified as four types, viz, P–P, P–G, G–G, G–J. As clearly seen in Fig. 1–15, P and G are pivotal major crops on which the other auxiliary minor crops are rotated. A set of crop-combinations expressed in the square bracket refer to multiple-cropping combination.

Crop rotations more than two sets of brackets which are connected with the symbol (=) could shift up a higher level of effective land use, viz, intensity of land use, thus leading to an increment of farm incomes.

Crop rotation in the brackets means double cropping whereby cultivator usually raises two crops per year on. An regards the average figure in the sample holding

28

Table 1-13 Compound effects of crop rotation on increasing fertility and productivity, Yerdona

Crop combination	Fertility increase in nitrogens (%)	Additional increase in yields (Q/acre)		
		Normal	Value realized	Gain
G – P	120 - 130	25	30	5
G – J	120 - 130	15 - 20	20	5
G – G	120 - 130			No gains

has been worked at 131.5%.

The favorable effects of crop rotation both on fertility and yields are manifold: (i) Crop rotation serves as a helping aids in controlling weeds, pests and diseases; (ii) It could make manure, midden composts and chemical fertilizer more effective by augmenting organic components of various kinds, which may exert beneficial effects on a succeeding crops; (iii) It reduces soil erosion and increases soil fertility by different crops.

Thus it follows that crop rotation results in additional increases in yields per acre. The empirical data, which has been collected from the very innovative farmer of the first rate with 65 acres and agricultural extension officer stationed in Yerdona, has ,been represented in Table 1-13.

The effects of crop rotation could be identifiable as compared with normal level without rotation. Level of nitrogen rises up practically as much as 20 to 30%. So does in yields except G–G combination. In the case of G–G combination there is a few discernible effects in fertility of crop rotation despite of the highest profitability than expected.

One of possible reason is considered to be a practical workability of so called rhizobium. This is attributable to the facts that rhizobium can not be active due to the water logging and salinization in the paddy fields, which is caused by inappropriate management of water supply and drainage.

Consequently, rhizome grown out in an excess supply of water tends to short lived. Farmer learned that leguminous crop as groundnuts has a power of feeding on free nitrogen of the air, while the legumious crop can draw only on the nitrogen supply stored in the soil.

Augmented amount of nitrogen by growing groundnuts is then ploughed back to the fertility stock needed for succeeding second crops which is usually consisted of fertility-consuming crops. Furthermore groundnuts outweights other crops except rice in profitabilities.

This is the reason why many farmers who have easy access to irrigation canal have so concerned with the growing groundnuts and selected it out as a pivotal crop in the crop combinations.

The existing several crop rotation types in which groundnuts plays a role of vital importance have deeply seated in agricultural production in Yerdona. And it should be noted that soil-augmenting effects of groundnuts can lead to lesser inputs of chemical fertilizer per acre. As will be discussed in the next section, fertilizer-saving effect has a far greater favorable consequences on a performance of agricultural management.

4.4 Productivities of Major Crops in Bidarakere and Yerdona

This section presents a brief analysis of productivities of major crops in the sample survey villages. The study of productivity per acre of crop leads on to an estimation of amount of products expressed in quintal per acre. Yields per acre mainly depends on the yields-seed ratio, degree of commercialization and current market price of crops concerned.

The degree of commercialization is dependent on the decision making of an allocation of agricultural produces among seeds, land rent, foodstuffs for livestocks and home consumption according to differential needs of each farms and marketabilities of crops and then the rest is to go on the market.

In this context productivities of major crop in term of yields-seeds ratio per acre and gross incomes per acre reveals an aggregated average performance of crop, with which carrying capacity of land under study is closely connected, and the degree of commercialization are primarily concerned with multivariate factors at work inherent in an infrastructure of each managements.

We have a few reports from those points of views based on a micro-level intensive field work. It is odd that it should have remained unexplored .The author, who has acutely been aware of the importance of how to ignite a potential of agricultural development and then lead it to a fuller fledged stage, carried out the intensive survey on the performance of crop-wise agricultural production. The crop wise statistical data are tabulated as both in Table 1-14 and Table 1-15 where from major findings have been deducted.

The first crop group, *ragi, jowar, huruli,* the second crop group, *thogari,* cotton and chili could be enumerated as a major traditional and commercial crops respectively in Bidarakere from a view point of the area

Table 1-14 Productivity of major crops in Bidarakere

Crop	House No.		1	2	3	4	5	6	7	8	9	10	11	12	13	14
			Varieties	Acreage		Gross production (Q)	Seed (kg)	Uses (Q)				Selling price (RS)	Productivity (Q/Acre)	Yields-Seeds ratio (4/5)	Gross income per acre (4×10/2)	Degrees of commercialization (9/4×100)
				Planted	Harvested			Livestock	Land rent	Consumption	Amounts sold					
	B-2	129-2		2		3	10			3	–	–	1.50	30.0	129.6	0
	B-3	122-2		1		1.5	6			1.5	–	–	1.50	25.0	129.6	0
	B-3	54		3.5		9	18			3	6	75	2.57	50.0	192.9	66.7
	B-3	66		4		12	24			6	6	80	3.00	50.0	240.0	50.0
	B-3	70		4.5		6	12			2	4	100	1.33	50.0	133.3	66.7
	B-3	10		3		2	20			2	–	–	0.67	10.0	57.6*	–
	B-1	119-1		3		5	15		2.5	2.5	–	–	1.67	33.3	144.0	–
	B-3	59		4		12	20			3	9	80	3.00	60.0	240.0	75.0
	B-3	45-1		4		12	15			12	–	–	3.00	80.0	259.1	0
	B-3	61	PR202	1.5		4	8			4	–	–	2.67	50.0	230.3	0
	B-3	9		2		12	10			12	–	–	6.00	120.0	518.2	0
	B-1	94		8		40	40			30	10	80	5.00	100.0	400.0	25.0
	B-3	113		4		15	12			–	15	80	3.75	125.0	300.0	100.0
	B-3	1		3		10	15			10	–	–	3.33	66.7	291.6	0
Ragi	B-1	17		10		36	35			–	36	81	3.60	102.9	490.0	100.0
	B-1	63		8		40	40			28	12	98	5.00	100.0	367.5	30.0
	B-1	101		16		60	50			30	30	80	3.75	120.0	300.0	50.0
	B-1	110	PR202	10		50	60			25	25	75	5.00	83.3	375.0	50.0
	B-1	109		12		50	60			25	25	85	4.17	83.3	351.2	50.0
	B-1	114		10		25	30			5	20	100	2.50	83.3	250.0	80.0
	B-2	37		10		45	50			25	20	90	4.50	90.0	405.0	44.4
	B-2	140		10		50	60			40	10	100	5.00	83.3	500.0	20.0
	B-2	59		10		65	100			45	20	70	6.50	65.0	455.0	30.8
	B-1	49		4		20	20			10	10	80	5.00	100.0	400.0	50.0
	B-2	67-3		20		50	80			20	30	100	2.50	62.5	250.0	60.0
	B-2	109		8		40	50			13	27	77.5	5.00	80.0	387.5	67.5
	B-2	47		6		40	30			20	20	102.0	6.67	133.3	680.0	50.0
	B-2	24		10		30	40			10	20	70	3.00	75.0	210.0	66.6
	B-3	19		10		30	40			15	15	100	3.00	75.0	300.0	50.0
	B-2	100		13		50	65			15	35	85	3.85	76.9	326.9	70.0
	B-1	102		4		20	8			–	20	98	5.00	250.0	490.0	100.0
	B-2	58	PR202	5		15	25			15	–	–	3.00	60.0	259.1	0
	B-3	10		2	0	0	5			–	–	–	–	–	–	–
	B-1	119-1	Hybrid	2		8	6			4	4	75	4.00	133.3	300.0	50.0
	B-3	59	Hybrid	1		3	1.5			1.5	–	–	3.00	200.0	257.8	0
	B-3	45-1		2		2	3			2	–	–	1.00	66.7	85.9	0
	B-3	61	CHS-5	3		2.5	7.5			2.5	–	–	0.83	128.9	71.6	0
	B-3	9		2		3	9			3	–	–	1.50	33.3	128.9	0
	B-1	94		3		15	6			15	–	–	5.00	250.0	429/6	0
	B-3	113		4		15	12			–	15	80	3.75	300.0	300.0	100.0
	B-3	1		1		5	3			5	–	–	5.00	166.7	429.5	0
	B-1	63		10		40	40			28	12	98	4.00	100.0	392.0	30.0
	B-1	110	CHS-1	10		30	30			30	–	–	3.00	100.0	257.8	0

30

Chap.1 RECENT TRENDS IN THE SOCIO-ECONOMIC STRUCTURE OF VILLAGES IN THE CENTRAL KARNATAKA

Crop																
Jowar	B-1	10		5		15	15		15		—		3.00	100.0	257.8	0
	B-1	11		9		30	27		20	10	85		3.33	111.1	283.3	33.3
	B-2	37		10		40	20		25	15	80		4.00	200.0	320.0	37.5
	B-2	140	CHS-5	9		20	25		20	—	—		2.22	80.0	190.9	0
	B-2	59	CHS-5	6		20	12		20	—	—		3.33	166.7	286.4	0
	B-1	49	CHS-5	3		15	6		13	2	100		5.00	250.0	500.0	13.3
	B-2	67-3		10		15	20		15	—	—		1.50	75.0	128.9	0
	B-2	105	CHS-1	12		60	36		20	40	76		5.00	166.7	380.0	66.6
	B-2	47	CHS-1	2		16	6		—	16	62		8.00	266.7	496.0	100.0
	B-2	24		20		50	50		20	30	70		2.50	100.0	175.0	60.0
	B-3	19	CHS-5	10		30	30		15	15	105		3.00	100.0	315.0	50.0
	B-2	100		13		100	39		25	75	100		7.69	256.4	769.2	75.0
	B-1	102		4		10	8		10	—	—		2.50	125.0	214.8	0
	B-2	58		6		20	12		10	10	100		3.33	166.7	333.0	50.0
Huruli	B-2	29-1		2		0.5	4	—	0.5	—	—		0.25	12.5		0
	B-3	66		4	0	0.5	8	0.5	0.5	—	—		0.13	6.3		0
	B-3	9		2		1	5	1		—	—		0.50	20.0		0
	B-3	113		4	0	1	3	1		—	—		0.25	33.3		0
	B-3	1		1		0.25	6	—	0.25	—	—		0.25	4.2		0
	B-1	17		10		0	9	—		—	—		—	—		0
	B-1	63		10		3	20	3		—	—		0.30	15.0		0
	B-1	101		2		3	10	3		—	—		1.50	30.0		0
	B-1	109		2		2	10	2		—	—		1.00	20.0		0
	B-1	114		4		4	15	4		—	—		1.00	26.7		0
	B-1	49		1		3	8	3		—	—		3.00	37.5		0
	B-2	67-3		10		3	10	3		—	—		0.30	30.0		0
	B-2	109		12		3	20	3		—	—		0.25	15.0		0
	B-2	24		10		5	5	5		—	—		0.50	100.0		0
	B-3	19		10		2	6	2		—	—		0.20	33.3		0
	B-2	100		13		6	50	6		—	—		0.40	12.0		0
	B-1	102		4		2	4	2		—	—		0.50	50.0		0
Thogari	B-3	9		2		0	10		0.5	—	—		—	—		— 0
	B-3	113		4		0.5	3		0.25	—	—		0.13	16.7		0
	B-3	1		1		0.25	5		0.25	—	—		0.25	5.0		— 0
	B-1	17		10		0	9		2	—	—		—	—		0
	B-1	63		10		2	6		0.25	—	—		0.20	33.3		0
	B-1	101		4		0.25	15		2	—	—		0.06	1.7		0
	B-1	109		1		2	15		1	—	—		2.00	13.3		0
	B-1	114		0.5		1	3		2	—	—		2.00	33.3		0
	B-2	37		1		2	30		2	—	—		2.00	6.7		0
	B-2	140		9		2	27		0.5	—	—		0.22	7.4		0
	B-2	67-3		10		0.5	10		1	—	—		0.05	5.0		0
	B-2	47		2		1	6		0.5	—	—		0.50	16.7		0
	B-3	19		10		0.5	10		0.5	—	—		0.05	5.0		0
	B-2	100		13		6	30		6	—	—		0.46	20.0		0
	B-1	102		4		0.5	8		0.5	—	—		0.13	6.25		0

(Continued)

31

Table 1-14 (Continued)

		1	2	3	4	5	6	7	8	9	10	11	12	13	14
			Acreage		Gross pro-duction (Q)	Seed (kg)	\multicolumn{4}{c}{Uses (Q)}	Selling price (RS)	Productivity (Q/Acre)	Yields-Seeds ratio (4/5)	Gross income per acre (4×10/2)	Degrees of commercialization (9/4×100)
Crop	House No.	Varieties	Planted	Harvested			Livestock	Land rent	Consumption	Amounts sold					
	B-3 84-1		2.75		4	10				4	390	1.45	40.0	567.3	100.0
	B-3 67		4		2.5	16				2.5	350	0.63	15.6	218.8	100.0
	B-1 119–1		1		0.75	2				0.75	300	0.75	37.5	225.0	100.0
	B-3 45–1		4		4	12				4	350	1.00	33.3	350.0	100.0
	B-3 9		2		2	9				2	300	2.00	22.2	300.0	100.0
	B-3 1		7		4	21				4	360	0.57	19.0	205.7	100.0
	B-1 17		10		10	25				10	330	1.00	40.0	330.0	100.0
	B-1 67		2		5	5				5	400	2.50	100.0	1,000.0	100.0
	B-1 101		8		8	20				8	350	1.00	40.0	350.0	100.0
	B-1 109		9		12	30				12	280	1.33	40.0	373.5	100.0
Cotton	B-1 114		8		6	24				6	380	0.75	25.0	285.0	100.0
	B-2 37		12		14	36				14	330	1.17	38.9	385.0	100.0
	B-2 140		10		10	25				10	300	1.00	40.0	300.0	100.0
	B-2 59		8		7	20				7	350	0.88	35.0	306.3	100.0
	B-1 49		6		12	9				12	352	2.00	133.3	704.0	100.0
	B-2 67–3		8		12	25				12	350	1.50	48.0	525.0	100.0
	B-2 109		6		4	18				4	315	0.67	22.2	210.0	100.0
	B-2 47		12		10	20				10	365	0.83	50.0	304.2	100.0
	B-2 24		6		7	12				7	300	1.17	58.3	350.0	100.0
	B-3 19		15		15	50				15	300	1.00	30.0	300.0	100.0
	B-2 100		25		25	75				25	350	1.00	33.3	350.0	100.0
	B-1 102		4		7	12				7	340	1.75	58.3	595.0	100.0
	B-3 84-1		2.75	0	0	8				–	–	–	–	–	–
	B-1 119–1		1	0	5	2			0.5	5	80	5.00	250.0	400.0	100.0
	B-3 9		2		0	5				–	–	–	–	–	–
	B-1 17		10		30	25				30	130	3.00	120.0	390.0	100.0
	B-1 63		2		6	4			0.5	5.5	80	3.00	150.0	240.0	91.6
	B-1 101		8		75	18			2	75	80	9.40	416.7	750.0	100.0
	B-1 110		5		12	12			1	10	140	2.40	100.0	336.0	83.3
	B-1 109		9		80	20			2	79	100	8.90	400.0	888.9	98.8
	B-1 114		8		75	?			1	73	160	9.40	?	1,500.0	97.3
Onion	B-2 37		8		75	20			1	74	150	9.40	375.0	1,406.3	98.7
	B-2 140		10		25	12			–	25	200	2.50	208.3	500.0	100.0
	B-2 59		4		40	10			1	39	40	10.00	400.0	400.0	97.5
	B-1 49		4		20	8			1	19	200	5.00	250.0	1,000.0	95.0
	B-2 67–3		8		30	15			1	29	170	3.80	200.0	637.5	96.7
	B-2 47		16		150	20			2	148	220	9.40	750.0	2,062.5	98.7
	B-2 24		6		150	12			2	148	200	25.00	1,250.0	500.0	98.7
	B-3 19		6		20	15			1	19	150	3.30	133.3	500.0	95.0
	B-2 100		25		75	16			2	73	200	3.00	468.8	600.0	97.3
	B-1 102		4		50	12			1	49	120	12.50	416.7	1,500.0	98.0
	B-3 84-1		2.75		1	0.5			0.5	0.5	400	0.36	200.0	145.5	50.0
	B-1 119–1		1		0.25	0.5			–	–	–	0.25	50.0	118.9	0

Chap.1 RECENT TRENDS IN THE SOCIO-ECONOMIC STRUCTURE OF VILLAGES IN THE CENTRAL KARNATAKA

	B-1	17	10		2	2	0.25		550	0.20	100.0	110.0	87.5
	B-1	63	2		0.5	0.5	0.5		–	0.25	100.0	111.9	0
	B-1	10	8		2	2	0.25	1.75	400	0.25	100.0	100.0	87.5
	B-1	110	3		4	3	0.5	1.75	500	1.33	133.3	666.7	87.5
	B-1	109	9		5	3	0.5	3.5	600	0.56	166.7	333.3	50.0
Chili	B-2	37	1		2	1	1	2.5	350	2.00	200.0	700.0	50.0
	B-2	59	4		1.5	–	0.5	1	450	0.38	–	168.8	66.7
	B-1	49	0.5		1	0.5	0.1	1	550	2.00	200.0	1,100.0	90.0
	B-2	67–3	8		2	–	0.5	0.9	500	0.25	–	125.0	75.0
	B-2	47	16		2	1.5	1	1.5	420	0.13	133.3	52.5	50.0
	B-2	24	6		1	1	1	1	–	0.17	100.0	74.6	0
	B-3	19	6		2	2	0.5	1	300	0.33	100.0	100.0	50.0
	B-2	100	25		10	5	0.5	9.5	350	0.40	200.0	140.0	95.0
	B-1	102	4		2	0.5	2	–	–	0.50	400.0	223.8	0

33

Table 1–15 Productivity of major crops in Yerdona

Crop		1	2	3	4	5	6	7	8	9	10	11	12	13	14
			Acreage		Gross production (Q)	Seed (kg)	Uses			Selling price (RS)	Productivity (Q/Acre)	Yields-Seeds ratio (4/5)	Gross income per acre (4 × 10/2)	Degrees of commercialization (9/4 × 100)
	House No.	Varieties	Planted	Harvested			Livestock	Land rent	Consumption	Amounts sold					
	B-5-132		0.30		6	30	–	–	6.00	–	–	8.0	20.0	731.4	0
	B-4-9		3.15		60	100	–	10	14.00	35	80	17.8	60.0	1,422.2	58.3
	B-2-80		4.00		60	120	–	–	58.80	–	–	15.0	50.0	1,371.3	0
	B-5-17		4.00	0	60	120	–	–	60.00	–	–	15.0	50.0	1,371.3	0
	B-5-157		1.00		8	20	–	–	5.80	2	85	8.0	40.0	680.0	25
			1.00		8	20	–	–	5.80	2	85	8.0			
	B-3-96		5.00		50	150	–	–	18.50	30	60	10.0	33.3	600.0	60
	B-1-48		12.00		167.75	375	–	–	13.25	150.75	72	14.0	44.7	1,006.5	89
	B-3-102		0.20		10	22	–	–	9.78	–	–	20.0	45.5	1,828.4	0
	B-5-100		6.00		100	120	–	–	23.80	75	100	16.7	83.3	1,666.7	75
Rice			6.00		100	120	–	–	–	98.8	100	16.7			98.8
(*Batta*)	B-3-49	White *Hamsa*	5.00		50	150	–	–	8.50	40	105	10.0	33.3	1,050.0	80
	B-3-78		3.00		45	120	–	–	6.30	37.5	75	15.0	37.5	1,125.0	83.3
	B-1-94		4.00		60	80	–	–	19.20	40	90	15.0	75.0	1,350.0	66.7
	B-1-8		4.00		60	75	–	–	14.25	45	100	15.0	80.0	1,500.0	75
	B-1-6		10.20		130	300	–	–	12.00	115	85	12.4	43.3	1,052.4	88.5
	B-3-99		6.00		135	50	–	–	19.50	115	100	22.5	270.0	1,956.3	85.2
			6.00		105	50	–	–	–	105	95	17.5	210.0		100.0
	B-1-79		34.00		450	850	–	–	61.50	380	95	13.2	54.0	1,282.5	84.4
			6.00		90	150	–	–	–	90	95	15.0			100.0
	B-3-30-1	*Masaru*	20.00		300	400	–	–	6.00	290	105	15.0	90.5	1,662.5	96.7
		White *Hamsa*	4.00		80	20	–	–	–	80	105	20.0			100.0
	B-1-22		6.00		67	300	–	–	27.00	37	105	11.2	22.3	1,172.5	55.2
	B-1-62		1.00		5	100	–	–	–	5	290	5.0	5.0	1,450.0	100.0
	B-4-9		3.15		15	400	–	–	–	15	220	4.4	3.75	977.8	100.0
	B-2-80		1.00		5	100	–	–	4.00	1	240	5.0	5.0	1,200.0	20.0
	B-3-104		1.00		10	70	–	–	–	10	310	10.0	14.3	2,530.0 (310.0)	100.0
			1.00		7	70	–	–	–	7	280	7.0	10.0	(1,960)	100.0
	B-1-63		1.20		4	150	–	–	–	4	280	2.7	2.6	746.7	100.0
	B-1-48		2.00		10	220	–	–	10.00	–	–	5.0	4.5	1,308.0	0
	B-3-41		8.00		40	480	–	–	–	40	220	5.0	8.3	1,100.0	100.0
	B-3-72		2.36		10	300	–	–	7.00	3	255	3.4	3.3	879.3	30.0
Ground nut	B-3-102		4.20		40	300	–	–	–	40	219	8.9	13.3	1,946.7	100.0
(*Nelagadale*)	B-5-100		6.00		15	600	–	–	2.00	28	275	5.0	25.0	1,375.0	93.3
					15	600	–	–	–	–	–		25.0		
	B-3-49		18.20		90	1,800	–	(1,700RS)	10.00	80	280	4.9	5.0	1,362.2	88.9
	B-3-78		4.25		35	300	–	–	3.00	32	300	7.6	11.7	2,270.3	91.4
	B-1-94		8.00		64	640	–	–	4.00	160	150	8.0	10.0	1,230.0	93.6
			12.00		100	960	–	–	–	–	–	8.3	10.4		

34

Chap.1 RECENT TRENDS IN THE SOCIO-ECONOMIC STRUCTURE OF VILLAGES IN THE CENTRAL KARNATAKA

B-1-8		4.00		30	450	–	–	–	30	250	7.5	6.7	1,875.0	100.0
B-1-6		8.00		30	120	–	–	–	30	225	3.8	25.0	843.8	100.0
B-3-99		18.00		189	1,800	–	–	25.00	164	270	10.5	10.5	2,689.6	86.8
B-1-79		8.00		70	800	–	–	–	70	270	8.8	87.5	1,477.8	100.0
		6.00		30	480	–	–	–	30	280	5.0	6.25	2,025.0	100.0
		30.00		160	2,400	–	–	–	160	280	5.3	6.67		100.0
B-3-30-1		20.00		125	1,500	–	–	30.00	95	280	6.3	7.81	2,025.0	76.0
		36.00		280	7,200	–	–	30.00	250	280	7.8	3.90		89.3
B-1-22		4.00		21	320	–	–	–	21	292	5.3	6.56	1,533.0	100.0
		4.00		31	480	–	–	–	31	263	7.8	6.49	2,399.9	100.0
		4.00		42	550	–	–	21.00	21	263	10.5	6.46		51.2
		10.00		52	800	–	–	–	52	268	5.2	6.5	1,393.6	100.0
B-1-62		1.00		11	4	–	–	3	8	70	11	275	770.0	72.7
B-2-83	Hyt=id	2.00		18	6	–	–	2	16	84	9	300	756.0	88.9
B-1-63		1.20		16	2	–	–	1	15	68	10.7	800	725.3	93.8
		4.00		0	4	–	–	–	–	–	–	–	*	–
B-2-60		6.00		4	10	–	–	4	–	–	0.7	40	57.1*	50.0
B-3-95	Hyt=id	3.00		30	6	–	–	15	15	60	10	500	600.0	0
B-5-110		5.00		5	10	–	–	5	–	–	1	50	85.6*	50.0
B-3-41		2.00		16	6	–	–	8	8	92	8	266.7	736.0	0
B-5-114		4.00		4	4	–	–	4	–	–	1	100	235.4*	0
B-3-72	Hyt=id	2.00		20	6	–	–	20	–	–	10	333.3	856.0*	0
		3.26		0	8	–	–	–	–	–	2.2	–	*	–
B-3-102		3.00		3	3	–	–	3	–	–	1	100	85.6*	0
B-3-73		4.00		4	8	–	–	8	–	–	1	50	85.6*	0
B-1-94		8.00		120	24	–	–	25	95	95	15	500	1,425.0	79.2
B-1-81		10.00		80	30	–	–	–	80	70	8	266.7	560.0	100.0
B-1-6		10.00		90	30	–	–	5	85	70	9	300	630.0	94.4
B-3-93		6.00		12	6	–	–	6	6	140	2	200	280.0*	50.0
B-3-33-1		15.00		12	15	–	–	4	8	125	0.8	80	100.0*	66.7
B-1-22	Hyt=id	10.00		150	30	–	–	10	140	68	15	500	1,020.0	93.3
B-2-83		0.50		4	3	–	–	4	–	–	2.7	133.3	–	0
B-2-42		8.00		0	10	–	–	–	–	–	–	–	–	–
B-3-41		6.00		18	12	–	–	9	9	60	2	150	180.0	50.0
B-1-8		10.00		20	15	–	–	–	20	60	2	133.3	120.0	100.0
B-1-77		9.00		10	10	–	–	10	–	–	1.1	100	–	0
B-1-22		4.00		30	80	–	–	30	–	–	7.5	37.5	–	0
B-1-63		4.00		0	4	–	–	–	–	–	–	–	–	–
B-5-157		0.10		1	3	–	–	1	–	–	4	33.3	–	0
B-5-110		9.00		4	8	–	–	8	–	–	0.4	50	–	0
B-3-41		8.00		24	8	–	–	12	12	80	3	30	240.0	50.0

Jowar (Jola)

Navane

Bajra (Sajje)

	1	2	3	4	5	6	7	8	9	10	11	12	13	14
		Acreage		Gross production (Q)	Seed (kg)	Uses			Amounts sold	Selling price (RS)	Productivity (Q/Acre)	Yields-Seeds ratio (4/5)	Gross income per acre (4 × 10/2)	Degrees of commercialization (9/4 × 100)
Crop	Varieties	Planted	Harvested			Livestock	Land rent	Consumption						
	House No.													
	B-3-104	4.00		2.5	30	–	–	–	2.5	300	0.63	8.3	187.5	100.0
	B-1-63	1.20		0.5	15	–	–	–	0.5	300	0.33	3.3	100.0	100.0
	B-2-69	4.00		–	20	–	–	–	–	–	–	–	*	–
	B-5-44	3.00		3/4	10	–	–	–	3/4	250	0.25	7.5	62.5	100.0
	B-3-72	10.00		0	30	–	–	–	–	–	–	–	*	–
Cotton	B-3-102	10.00		2	50	–	–	–	2	250	0.2	4.0	50.0	100.0
(*Hatthi*)	B-5-100	4.00	*Jaidar*	20	5	–	–	–	20	400	5	400	2,000.0	100.0
	B-3-78	12.00		6	60	–	–	–	6	300	0.5	10	150.0	100.0
	B-1-94	6.00		0	–	–	–	–	–	–	*	–	*	–
	B-1-6	7.00		3	20	–	–	–	3	315	0.43	15	135.0	100.0
	B-3-99	6.00		6	6	–	–	–	6	250	1	10	250.0	100.0
	B-3-30-1	5.00		15	5	–	–	–	15	400	3	300	1,200.0	100.0
Horse gram (*Huruli*)	B-5-44	7.00		3	25	2	–	1	–	–	–	–	–	0
Bengal gram (*Kadale*)	B-3-102	1.00		2	12	–	–	2	–	–	–	–	–	100.0
	B-3-78	1/2		1	6	–	–	1	–	–	–	–	–	100.0
Maize (*Muskina jola*)	B-3-99	6.00		48	24	–	–	–	48	100	–	–	800.0	100.0
Sugarcane (*Kabbu*)	B-3-96	2.00		15	6	–	–	–	12	140	–	–	1,050.0	80.0

* Crop failure

Table 1–16 Crop-wise productivity, selling price and gross income per acre of the sample villages

Bidarakere	Productivity (Q/A)			Selling price (Rs)			Gross income per acre (Rs/A)		
	χ	σ	c.v.	χ	σ	c.v.	χ	σ	c.v.
Ragi	3.63	1.52	41.87	86.37	10.88	12.60	322.8	131.0	40.58
Jowar	3.56	1.79	50.28	85.92	14.20	16.53	304.3	154.0	50.6
Cotton	1.18	0.50	43.47	338.27	32.24	9.53	387.9	188.4	48.6
Onion	7.35	5.64	76.73	142.35	53.09	37.30	800.7	521.1	65.1
Chili	0.59	0.62	105.71	447.50	93.33	20.86	266.9	296.4	111.0
			63.6			19.4			
Yerdona	Productivity (Q/A)			Selling price (Rs)			Gross income per acre (Rs/A)		
	χ	σ	c.v.	χ	σ	c.v.	χ	σ	c.v.
Rice	14.39	3.93	27.31	91.42	12.83	14.03	1,268.30	381.96	30.12
Groundnuts	6.44	2.17	33.69	261.60	34.16	13.06	1,553.03	570.16	36.71
Jowar	10.57	2.54	24.03	85.64	25.78	30.10	807.00	253.00	31.35
Cotton	0.37	0.16	43.24	307.22	58.37	18.99	–	–	–
			32.1			19.0			

covered. In Yerdona, there could be counted that rice, groundnuts and sugarcane turn out to be commercial crops and then *jowar, navane, bajra* and pulses of various kinds come out as traditional one.

As for the degrees of commercialization in Bidarakere, cotton comes first and is put on the market completely, followed by onion, chili. *thogari* and *huruli* are usually given to the livestock as feedstuffs. None of them in the sample holdings are marketed.

They are also grown for the purpose of maintaining soil fertility as a mixed crop, mainly being combined with traditional crops for a staple as *ragi, jowar*. Thus the degrees of commercialization of those crops decrease considerably with a varying proportion from 25% to 100%. It depends both on the amount of basic needs for staple and the quantities produced in each holdings. As compared with Bidarakere, the degrees of commercialization of eleven crops there could be seen some identifiable trends in varying degrees. Groundnuts and cotton are completely sold out at a comparatively brisk rate. Bengal gram and sugarcane are also marketed but the samples are so limited in its number that it can not be clear. Unlike the case of *jowar* in Bidarakere, where it grows mainly for one of the staples, degrees of commercialization are counted as high as 50% to 100%, except crop failure cases.

Traditional crops as *navane* and *bajra* growing on a drier and eight irrigation lands mostly tend to swing extremely from zero to a hundred percent due to the availability of irrigation water. The crops on the complete dry land are likely to be damaged by a drought, thus resulting in crop failure.

Lastly, as for rice, the bulk of it tends to be sold, however, the amount sold is largely dependent on a particulars of demand and supply in each holdings. These shows general tendency that the ratio of the amount sold to total production is decreasing in inverse proportion to the size of agricultural holdings. It indicates how much rice as a staple is needed in each specific households. In Yerdona, the staple foods are mainly composed of rice regardless of holding size. The people who can afford to take a rice every day as a staple is extremely confined to the high class.

Principal findings which have been reduced from empirical data are succinctly shown in Table 1–16 and at the same time, data has been dipicted graphically both on Fig. 1–17 and Fig. 1–18.

Fig. 1–17 shows correlation of crop-wise yields per acre (Q/A) and selling price expressed in coefficient value (C.V.) of Yerdona. Distribution of the four major crops of Yerdona relatively clusters in a group as compared with that of Bidarakere. Moreover, unlike selling price, C.V. of yields per acre are widely distributed as for Bidarakere with average percentage of 63.6, two times greater than that of Yerdona. On the other hand, selling prices of each crop groups are levelling off at about the level of 19%. The widely scattered C.V. of crop-wise productivity per acre could be partly ascribable to the commercial crops as *chili* and onion, however, the dispersion within each crop group are also comparatively greater than that of Yerdona in varying degrees. The C.V. of *ragi* and cotton in Bidarakere and *jowar* and rice in Yerdona tend to be much more stable with respect to productivity per acre.

In summing up those differential trends clearly seen in productivity and selling price indicate the fact that productivity per acre among the different holding size varies in a wider range, because compounded effects both of locational conditions as soil fertility, availability of water needed for growing a crop and entreprenencial efforts are clearly reflected in the marked differences in gross income per acre.

Further, it must be added that widely dispersed pattern of productivity (Q/A) is well represented in the different levels of a capital coefficient (Y/K). In other words, permitting that coefficient level of each holding

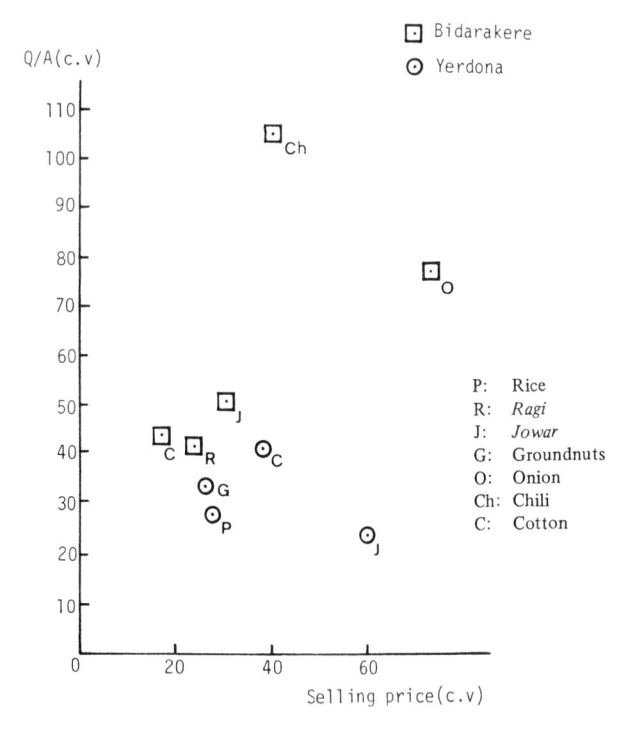

Fig. 1–17 Relation between coefficient values of productivity (Q/A) and selling prices of major crops

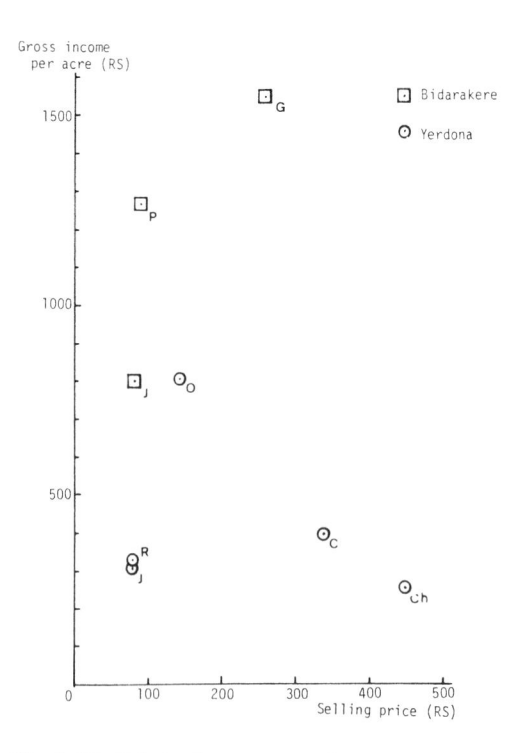

Fig. 1–18 Relation between gross incomes per acre and sellingof proces of major crops

is on the same levels ; and other condition being equal, productivity per acre expressed in C.V. tends to get closer and converge on a same level. Differences in a capital coefficient is directly connented with technological gap in agricultural production.

When it comes to the gross income per acre (G.I.) it varies widely from 200 rupees level to 800 rupees level in Bidarakere, and it jumps up to 1,200 rupees level in yerdona. As the C.V. of selling price is levelling off at about 19% level, the difference of G.I. within a crop and among various crops could be largely attributed to the productivities (Q/A) of each crops and holdings. This point will be discussed in the next section with a reference to an identifiable trends in a comparative productivities of gross farm income per acre.

5. Farming Size and Economic Performances of Farm Management

Statistical particulars and details pertaining to the economic performances of farm management in Bidarakere and Yerdona have been prepared not only for clarifying a salient features of framework relevant to agricultural production, but making a rough estimation of pay-off balance between inputs and output. Peacemiel data relevant to agricultural production has been collected by applying the schedule to sample holdings of which results was critically scrutinized by the intensive hearing during the field survey. And then has been as-

sorted into the three categories, viz, family size, pay-off balance of inputs and outputs and living costs and loans. The results are tabulated both in Table 1–17 and Table 1–18.

One of the most annoying problems rests on the interviewing of delicate nature. A greater part of farmers are more reluctant to be interviewed. The interviewing process and follow-up checking of the results might be an intriging combination of causes and effort, in which intrinsic maltivariate forces are at work on shaping pay-off matrix patterns. But this could only a time consuming and interferring vexation to the interviewees.

The author has tried not to step on farmers corner right and left with the sensitivity of an ox as much as possible. This attitude toward farmers would lead to chanelling farmer's interest to us, then a set of reliable informations to the interviewer.

Major fact findings could be summed up as follows:

(1) As for Bidarakere, to begin with the indicators agricultural holding and livestocks, which could be bench marks of farming size the bulk of sample holdings runs unirrigated dry land cultivators of wet land mainly composed of extremely elongated paddy fields are concentrated in the higher class with the acreage of well over 20 unirrigated land. The holding of B–1–119 is the exceptional case.

(2) Stock holding size tends to grow as land holding size moves upward. Small livestocks show a tendency to be concentrated in the marginal and middle holding classes. By contrast, big livestocks tend to be

concentrated to the bigger farm.

This is closely relevant to the means of production in labor process. Almost all farms run their farms depending solely on draft cattle and manual agri-cultural implements instead of motors and working machines. Marginal farmer can hardly afford to buy expensive pair of draft animals. When purchased, a pair of bullocks as a draft animals cost at 2,000 to 3,000 rupees.

This amount of money nearly corresponds to a yearly gross income. This naturally leads them to hiring a *badige* instead. *Badige* system is composed of a pair of bullocks, their drover and labour gang who leads a helping hands to him and goes around fields to be hired on reason. They receive money or agricultural produces in kind for their services. This is the reason why *badige* is mostly concentrated in the marginal farmer class.

(3) As regards inputs, out of outlays that has been spent on each items, payments to hired labourer, fertilizer, seeds, young plants and transportation charges tend to have a higher proportion as compared with other items. On the average, the current costs are far greater than the fixed costs.

Farmers in Bidarakere have been increasingly burdened with financial deadweight. Most of farmers have borrowed an enormous amount of money needed mainly for the fertilizer seeds and wages pay-offs with amazingly high interest rated 20 percent per annum from money lenders. Qualification for the loans with varying interest rates less than 20 percent is substantially limited to the wealthy class in their solvency.

A handful farmer make the most use of bank loan rendered by governmental bodies or public sector. The rest has no alternative but unwillingly go to money lender who lends the amounts as he wants on the long term and squeeze out completely. It follows then that overloan exceeding far over his solvency shall plunge him into a desolated state in the long run.

These trends as described above, combined with subdivision of land caused both by a growing tendency of individualism among the co-heirs in a patriarchally extended or jointed family and time honored equal share division inheritance system, eventually has an unfavorable effects on an economic pay-off balance of farm management.

They make a meagre living in a reserved manners with a monthly living expenses in varying degrees. The households who usually spend living expenses over 500 rupees per month accounts only at seven out of thirty-five households. The picture we have seen from the economic point of view is ever so bleak. Comparative examination of productivity per acre shall reveal considerable performance gaps in some respects between the sample survey villages, Bidarakere and Yerdona.

(1) In Yerdona, the ratio of the irrigated land to the unirrigated land stands considerably high in compared with that of Bidarakere, almost the reverse is the case. In addition, the holding who cultivates the leased-in land amounts to 55 percent, which is nearly 2.8 times greater than that of Bidarakere.

The most of the holding have bullocks more than two sets as a draw animal. Cows yielding milk for domestic uses are evenly distributed across the different holding size classes. The same is true for Bidarakere.

(2) As regards gross farm income, it can not be disregarded that productivity level is so high that a greater part of holding tends to make a living depending mainly on agricultural incomes.

On the other hand, the ratio of dependence on the permanent agricultural labour income arises, according as the holding size goes down, and particularly when the ratio of the dry land to the irrigated land increases. This could be true of even the case of B–5–140 holding composing totally of the unirrigated land, 14 acres in extent. While admitting that the crops grown have been completely damaged by the shortage of water and the severe drought, resulting in crop failure. The gap of economic performances between a dry land and an irrigated land turned out to be even far greater than expected.

Low productivity of land leading thus to low income tends to be supplemented by permanent or casual agricultural labor income in some ways. This tendency is particularly magnified in Bidarakere where the most farmers could only depend on the infertile dry land.

It is interesting to note that critical threshold class that farmer can barely make a living by a meagre income from dry land and an additional laborer's wages of family member lies around the farming size with less than 10 acres in both sample villages, below which they have narrowly enough food to sustain life without earning agricultural wages.

(3) The amount of inputs per acre involved in farm expenditure is considerably high in compared with that of Bidarakere. Costs of chemical fertilizer, insecticide, herbicide, seeds and young plants have a high proportion as well as wage payments to hired laborer. Cost for irrigation water including maintenance cost and land tax can not be disregarded.

(4) As farmers become involved in the agricultural market, they naturally tend to be profit-oriented. The pressure of ever-increasing prices particularly of fertilizer, seeds, young plants and wages drive them the indebtedness.

As the result, total balance of financial position is generally going into the red across the sample holding. 14 holdings out of 25 sample holdings are indebted to many lenders and banks for a large sum. As interest rate is so high varying from 11 to 48 percent per annum, their debts are not to be discharged easily.

Table 1–17 Farming size and economic performance of farm management in Bidarakere

		I								II					III		
	Household No.	B-3-108	B-2-29-2	B-3-122-2	B-3-84-1	B-3-54	B-3-66	B-2-67-3	B-3-70	B-3-10	B-1-119-1	B-3-59	B-3-61	B-3-9	B-1-94	B-3-113	B-3-45-1
	Household members	4	1	8	7	8	7	2	10	9	6	4	7	7	11	9	10
Farming size	Agricultural holding																
	Irrigated land																
	Owned										1.20						
	Leased-in																
	Leased-out																
	Unirrigated land																
	Owned	1.30	2.00	2.00	2.30	3.20	4.00	4.00	2.20	5.00	1.20	5.00	8.20	8.20	11.0	12.00	7.00
	Leased-in								2.00		3.00	2.00					5.00
	Leased-out																
	Total holding	1.30(10)	2.00(10)	2.00(10)	2.30(10)	3.20(15)	4.00(10)	4.00(20)	4.20(20)	5.00(10)	6.00(25)	7.00(40)	8.20(25)	8.20(12)	11.00(20)	12.00(12)	12.00(25)
	Livestocks																
	He-Baffalo																
	She-Baffalo														5(2)	3	
	Bullock						2		4(2)	2	2			2		2	
	Ox																
	Cow			1		3	4		1	1	2(1)					1	3(1)
	He-Goat								1								
	She-Goat																
	Sheep																
	Cock													1			
	Hen		2		4		14				2	1					15
Gross income	Agriculture	450	300	–	2,500	1,275	1,000	800	600	1,000	5,000	840	1,000	2,500	4,000	2,500	2,105
	Cattle breeding																
	Land rent																
	Casual agr. labour	1,800	150	2,060	200	2,060		200	1,000	500		200	401				
	Permanent agr. labour																
	Business						1,200		730					100			
	Working away from home																
	Others				6,960								3,600			8,400	
	Total	2,250	450	2,060	9,660	3,335	2,200	1,000	2,330	1,500	5,000	1,040	5,001	2,600	4,000	10,900	2,105
Farm expenditures	Chemical fertilizer	30	200	30	300	300	280		150	200		300	800	500	400	200	1,000
	Organic fertilizer	500	200		250	300	300		200	100	300	120	400	500	300	500	500
	Insecticide, herbicide												100				
	Seeds and young plants	30	15	9	250	70		32	20	150	300	100	360	200	36		200
	Animal feeds								75		100						100
	Fuel, electricity										350						
	Payments to hired labourers		400		550	100	250	550	250		4,109	200	700	300	500	300	350
	Land rent								400		185						
	Costs for irrigation water										40						
	Transportation charge	10			400	150	50	100			15	100	300	100			200
	Purchase of small animals									25	100			50			200
	Purchase of large animals								1,250								500
	Purchase and repairments of farming tools or machines										500						
	Maintenance of irrigation																
	Land tax	3	6	5	10	5	10	8	6	9	60	20	15	16	15	20	40
	Other farm expenditures			100		350							100				200
	Total farm expenditures	573	821	144	1,760	1,275	890	690	2,351	484	6,059	840	2,775	1,666	1,251	1,020	3,290
		Badige	Badige	Badige		Badige	Badige	Badige									
Living expenses and loan	Monthly living expenses of household	100	50	300	500	300	300	250	150	100	100	200	250	50	150	100	350
	Loan																
	Amount	300	–	25	–	4,000	300	–	–	–	–	400	2,000	500	–	–	–
	Interest rate p.a.(%)	24				24						36	12	18			
	Purpose																
	(1)					Marriage (2,500)	Fertilzer					Fertilizer		Fertilizer			
	(2)					Other (1,500)						Seeds		Seeds			
	(3)																
	Gross incomes par Capita	562.5	450	257.5	1,380	416.9	314	500	233	166.7	833.3	260	714.4	371.4	363.6	1,211.1	210.5
	Gross farm income per acre	257.1	150	–	909	364.3	250	200	133.3	200	833.3	120	117.6	294.1	363.6	208.3	175.4

40

Chap.1 RECENT TRENDS IN THE SOCIO-ECONOMIC STRUCTURE OF VILLAGES IN THE CENTRAL KARNATAKA

			IV		V								VI					
B-3-1	B-1-102	B-1-17	B-1-63	B-1-101	B-1-110	B-2-109	B-1-114	B-2-37	B-2-140	B-2-59	B-1-49	B-3-67-3	B-1-109	B-2-47	B-2-24	B-3-19	B-2-100	B2-58
6	11	7	13	8	14	6	9	7	9	13	10	7	6	7	12	9	15	8
						2.20			1.20	8.00	4.00					1.02	1.10	
14.00	16.20	14.00	23.00	27.00	27.00	27.20	30.00	28.00	34.12	27.00	31.00	39.37	40.00	46.00	47.00	43.38	54.38	37.00
		5.00			3.00			3.00										26.00
14.00(14)	16.20(30)	19.00(25)	23.00(50)	27.00(35)	30.00(20)	30.00(40)	30.00(30)	31.00(30)	35.32(35)	35.00(35)	35.00(35)	39.37(40)	40.00(40)	46.00(20)	47.00(47)	45.00(45)	56.08(57)	63.00(40)
3																		
	4(4)	2(1)		5(1)	5(2)		5(3)	5(2)	2(1)	1(1)	3(1)	1	2(1)	3(2)	2(2)			
	2	2	4	6	4		4	4	4	4	2	4	4	4	4	4		
																1		
		2(1)	5(2)	4(1)		4(2)	3(1)		16(4)	8(4)	4(2)	1(2)	3(3)	5(3)	8(4)	6		2
										2						2		
2									2						17			
									8									
3,000	15,000	11,078	9,000	11,900	4,025	17,000	17,570	19,070	16,000	6,210	20,000	17,430	18,000	45,144	35,200	15,000	37,150	3,500
																		960
																	5,000	
									7,200									720
3,000	15,000	11,078	9,000	11,900	4,025	17,000	17,570	19,070	23,200	6,210	20,000	17,430	18,000	45,144	35,200	15,000	42,150	5,180
1,500	1,500	1,500	1,200	2,000	2,000	3,000	3,000	4,000	5,500	3,000	1,000	3,000	3,000	2,500	3,000	5,000	6,000	1,000
200	1,500	900	1,000	1,000		2,000	2,000	1,000	2,500	2,000	1,000	1,000	1,500	1,300	2,000	1,000	3,000	
				100	200	500	200	600	200	200	500		500		300		400	
150	1,500	1,000	300	1,300	300	1,000	3,000	2,500	1,000	1,000		1,000	1,500	2,320	600	1,500	2,000	
	250	125		500	400	500	200	500	800	200		300	1,000		1,000	500	500	
								1,000	400	400							180	
600	2,500	2,500	1,000	3,000	4,000	5,000	3,000	8,000	3,500	4,000	3,500	3,000	3,000	5,000	10,000	3,000	10,000	2,500
						35			50							150		300
50	1,500	200	300	1,500	2,000	2,000	1,000	1,500	800	2,000		1,000	500	3,000	1,000	1,000	2,000	
	200		100		100	300	200	500	200	300	1,500	200	300	100		100	500	
			1,235	1,800				3,500			2,800	560	2,000	2,800	3,000		1,300	
					338	300		1,000	250	200				200				
			50			500												
30	25	30	40	55	300	40	800	200	42	30	350	150	100	300	50	150	180	40
	500		50						1,100			200		500	1,000			
2,530	9,475	8,055	5,275	11,255	9,638	16,175	13,200	23,300	16,342	13,330	10,650	10,410	13,400	17,520	21,450	13,400	26,060	3,840
Badige																		
100	200	200	150	300	200	500	500	250	300	200	500	300	600	350	600	250	600	350
–	–	1,500		4,000	10,000	2,250		3,000	4,000	4,000	1,500		5,000		5,000	1,500	5,000	
		15		15	14	14		27	24	14	18		12		12	12	15	
		Fertilizer		Fertilizer	Fertilizer	Fertilizer		Fertilizer	Fertilizer	Pranting of coconuts	Fertilizer		Agriculture		Agriculture	Fertilizer	Fertilizer	
		Seeds			Seeds	Seeds		Seeds	Seeds				Seeds Wage		Fertilizer Seeds	Wage		
500	1,363.6	1,582.6	692.3	1,487.5	287.5	2,833.3	1,952.2	2,724.3	2,577.8	477.7	2,000	2,490	3,000	6,449.1	2,933.3	1,666.7	2,810	647.5
214.3	909.1	583.1	391.3	440	134.2	566.7	583.3	615.2	446.9	177.4	571.4	436.6	450	981.4	748.9	333.3	661.0	70.8

41

Table 1–18 Farming size and economic performance of farm management in Yerdona

		I					II				III			
	Household No.	B-4-132	B-1-62	B-4-9	B-2-80	B-5-17	B-3-104	B-1-63	B-5-157	B-2-42	B-2-69	B-3-96	B-5-140	B-1-48
	Household members	6	4	10	5	9	4	8	10	7	4	11	10	13
Farming size	Agricultural holding													
	Irrigated land													
	Owned	0.30	1.00	1.35	3.20	4.00	1.00	1.20	1.10			10.00		14.00
	Leased-in			1.20										
	Leased-out													
	Unirrigated land													
	Owned						4.00	5.20		8.00	10.00		14.00	
	Leased-in													
	Leased-out								6.25					
	Total holding	0.30(3)	1.00(3)	3.15(10)	3.20(10)	4.00(5)	5.00(4)	7.00(8)	7.35(5)	8.00(5)	10.00(10)	10.00(40)	14.00(10)	14.00(5)
	Livestocks													
	He-Baffalo											1(1)		
	She-Baffalo	2		2					1					1
	Bullock		2	2	2				2		4	1		6
	Ox													
	Cow	1		1	2			1		2	1	2(2)		1(1)
	He-Goat													
	She-Goat													
	Sheep													
	Cock													
	Hen													
Gross income	Agriculture	*	2,010	6,100	1,584	7,000*	3,750	2,000	1,500	**	**	4,660	*	20,000
	Cattle breeding													
	Land rent													
	Casual agr. labour	3,650	1,090	900				3,300	1,000	7,500	1,500		5,375	
	Permanent agr. labour													
	Business	330					4,000	270						
	Working away from home													
	Others													
	Total	3,980	3,100	7,000	1,584	7,000	7,750	5,570	2,500	7,500	1,500	4,660	5,375	20,000
Farm expenditures	Chemical fertilizer	300	300	2,000	2,500	5,000	200	1,000	600			6,500		10,000
	Organic fertilizer		100	400	100		300		100	60	100	400	200	1,000
	Insecticide, herbicide			200	400		100	150	20			2,000		500
	Seeds and young plants	40	450	2,000	60		400	1,200	24	20	70	2,500		
	Animal feeds		100		64					60		1,000		
	Fuel, electricity													
	Payments to hired laborers	300		1,000	1,500	1,500	1,200	600*	150	125	300*	6,000	1,000	8,000
	Land rent			914										
	Costs for irrigation water	20	75	150	100		75	120				1,100		1,010
	Transportation charge		100	300	100		80	200	20			200		250
	Purchase of small animals													
	Purchase of large animals													
	Purchase and repairments of farming tools or machines											500		
	Maintenance of irrigation		10	200			30	150				500		600
	Land tax	100	50	40	40	–	65	30	120	15	15			420
	Other farm expenditures												30	
	Total farm expenditures	760	1,185	7,004	4,664	6,900	2,450	3,450	1,034	280	485	20,700	1,230	21,780
							Badige				*Badige*			
Living expenses and loan	Monthly living expenses of household	600	150	300	100	700	200	200	150	450	150	600	400	500
	Loan													
	Amount		500		4,000			2,000				7,000		25,000
	Interest rate p.a.(%)		24		24			24				12.5		24
	Purposes													
	(1)		Seeds		Agriculture			Seeds				Agriculture		Agriculture
	(2)		Fertilizer		Fertilizer			Fertilizer				Dowry		
	(3)													
	Gross farm incomes per acre	–	2,010	1,807.4	452.6	1,750	750	285.7	190.5	–	–	466		1,428.6
	Gross incomes par Capita	793.3	775	700	316.8	777.8	1,937.5	696.3	250	1,071.4	375	423.6	537.5	1,538.5

Notes: Figures in the brackets succeeding the column of total holding are supposed to be an ideal agricultural holding size enough to maintain household by H.H.
  * Agricultural products are used mainly for home consumption.
  ** Crop failure

Chap.1 RECENT TRENDS IN THE SOCIO-ECONOMIC STRUCTURE OF VILLAGES IN THE CENTRAL KARNATAKA

	III					IV			V			VI		
	B-3-41	B-5-100	B-5-144	B-3-72	B-3-102	B-3-49	B-3-78	B-1-94	B-1-8	B-1-6	B-3-99	B-1-79	B-3-30-1	B-1-22
	9	9	10	15	7	11	6	11	6	9	8	10	7	12
	10.00	16.00	11.00	4.36	19.00		8.00	20.00	32.00	35.20	30.00	37.00	50.00	35.28
	6.00		7.00	13.26		15.00 8.20	16.25	6.00			6.00	9.00	10.00	27.00
	16.00(16)	16.00(16)	18.00(28)	18.22(25)	19.00(20)	23.20(25)	24.25(25)	26.00(26)	32.00(32)	35.20(34)	36.00(36)	46.00(46)	60.00(100)	65.28(75)
	2	3		1(2)	3	2(2)	1	4	4	4	5	6	4(2)	7
	4	2		2	3	4	4	4	4	4	6	5	6	4
	2	4	4	1(1)	4	2(2)			1		4	3	6(3)	2
						5 25								
	11,036	33,200	15,000	13,000	11,000	26,600	18,000	36,625	18,800	35,000	93,195	97,850	127,450	100,000
			6,000								10,000			
											20,000			10,000
	11,036	33,200	21,000	13,000	11,000	26,600	18,000	36,625	18,800	35,000	123,195	97,850	127,450	110,000
	2,000	12,000	400	4,000	1,000	6,000	4,000	3,000	4,600	15,000	13,000	15,000	12,000	10,000
	600	1,000	100	400	2,000	800	500	700	600	2,000	1,800	2,000	600	2,500
		8,000		1,000	400	1,500	2,000	200	200	3,000	2,000	800	3,000	3,000
	100	1,500		2,500	1,500	240		2,000	2,000	10,000	10,000	15,000	5,000	13,000
	300	1,000		300					1,000	1,500	1,200	600	3,000	800
											1,200			2,800
	6,400	16,000	600	2,000	7,000	3,000 1,700	8,000	13,000	10,000	20,000	9,000	25,000	24,000	26,000
	750	1,200		750	200	800	1,000	1,000	1,500	2,000	2,250	3,375	5,000	2,663
	200	1,000	30	100	150	250	300	400	1,000	2,000	1,000	4,000		2,000
		200		300					100	200	1,000	5,000		
									3,200	880			5,000	
													10,000	3,700
		250		100	60	200	300		200	500	3,000	4,000		900
	600	500	2,400 (9+16)	600	50 600	800		800	600	5,000	900	4,000	5,000	7,556
	10,950	42,650	3,530	12,050	12,960	15,290	16,100	20,600	25,000	62,080	46,350	78,775	72,600	74,919
			Badige		Badige									
	500	1,000	500	1,000	1,000	500	500	400	500	1,000	300	1,000	1,700	100
	3,000	26,000		7,000		7,000		10,000	20,000	20,000			15,000	2,500
	48	20,000(18) 6,000(12)		48		18		24	24	24			11	11.5
	Seeds Fertilizer	Seeds Fertilizer						Seeds Fertilizer	Seeds Fertilizer	Seeds Fertilizer			Seeds Fertilizer Wages	Seeds Fertilizer
	689.8	2,075		700.8	578.9	1,131.9	731.0	1,408.7	587.5	985.9	2,588.8	2,127.2	2,124.2	1,522.0
	1,266.2	3,688.9	2,100	866.7	1,571.4	2,418.2	3,000	3,329.5	3,133.3	5,888.9	15,399.4	9,785.0	18,207.1	9,166.7

Table 1-19 Gross farm income per acre and gross income per capita of the sample villages

(Unit=RS)

Landholding size		G.F.I.* per acre Yerdona ($\bar{\chi}a$)	G.F.I.* per acre Bidarakere ($\bar{\chi}b$)	Productivity gap ($\frac{\bar{\chi}a}{\bar{\chi}b}$)	G.I.** per capita Yerdona ($\bar{\chi}c$)	G.I.** per capita Bidarakere ($\bar{\chi}d$)	Productivity gap ($\frac{\bar{\chi}c}{\bar{\chi}d}$)	Economic classification of farming size Yerdona	Economic classification of farming size Bidarakere
I	0 ~ 4.9	1,505.0	323.4	4.65	672.6	514.2	1.31	} marginal	} marginal
II	5.0 ~ 9.9	408.7	313.0	1.31	988.8	469.2	2.11		
III	10.0 ~ 19.9	989.8	409.3	2.42	1,374.2	871.9	1.58	} lower middle	} lower middle
IV	20.0 ~ 29.9	1,090.5	415.7	2.62	2,915.9	1,089.9	2.68		} upper middle
V	30.0 ~ 39.9	1,387.4	441.5	3.14	7,473.9	1,917.9	3.90	} upper middle	} large
VI	> 40	1,924.5	634.9	3.03	12,386.3	3,371.8	3.67	} large	
$\bar{\chi}$		1,199.65	418.79	2.86	3,259.78	1,355.08	2.41		
S.D.		711.51	254.67		4,582.89	1,317.79			
C.V.(%)		59.31	60.81		140.59	97.25			
Linear regression equation		y=131.47+ 324.91$\bar{\chi}$ (r^2=0.953) excl.I	Y=238.81+ 67.69$\bar{\chi}$ (r^2=0.822) excl.I						

* Gross farm income   ** Gross income

They are now torn between the repayment of their debt and in increasing additional costs of inputs under the inflationary pressure. One of the most difficult problems now vexing many farmers is to run a farm just to keep from going into bankrupcy. The problem is still in the balance.

(5) The sample holding is increasingly polarized into the affluent class and the impoverished one. Dividing ridge is assumed to be the lower and upper middle class ranging from 15 to 35 in acreage.

In conclusion. the facts on the relation between farming size and economic performance of farm management in the sample holdings could be summarized as in Table 1–19.

(1) The average gross farm income per acre in Yerdona and Bidarakere is estimated at 1,199.65 rupees and 418.79 rupees respectively, being 2.89 times as high as in Yerdonaas in Bidarakere. In other words, G.F.I., per acre in Bidarakere averaged only 34.9% of that in Yerdona. Roughly speaking, the productivity of dry land of one acre is equivalent to the irrigated land of ca 2.9 acres in the monetary term. The maximum productivity gap between two sample villages can be seen in the lowest class up to 4.9 acres, and minimum gap in the second lowest class ranging from 5 to 9.9 acres.

This could well explain not only by the facts that ratio of the dry land to irrigated land in the II class of Yerdona is so high that the productivity is to decline drastically, but the absolute holding size itself is limited in acreage in the lowest class of Yerdona, it follows thus the marginal farmers are compelled to make the most use of it as much as possible. The organized efforts compelled by the necessity to make a living with their land of limited size urge them to intensify land use by the improved multi-cropping farming methods.

(2) G.I. per capita of Yerdona is 2.41 times greater than that of Bidarakere. G.I. gap between two samples is widening in proportion to farming size. But the G.I. of Yerdona tends to increase at the exponential rate, being 1.47 for II, 2.04 for III, 4.33 for IV, 11.11 for V and 18.42 times for VI respectively as compared with class I (1.00).

There is a marked difference between class IV and VII. By contrast, in Bidarakere it increases at a moderate rate with some drop in the class II, being 1.70 for III, 2.12 for IV, 3.73 for V and 6.56 times for VI respectively as compared with class I (100).

(3) In summing up these facts as have discussed above, the economic farming size could be classified as in Table 1–19 , judging from economic performances of farm management which is well reflected in the productivity per acre in monetary term.

6. A New Frontier and "Irrigation Rush" in Yerdona

Yerdona village is situated in the southern part of Deccan plateau, second driest semi-arid zone next to the Thar. Yerdona has long been vexed with the recurrent drought with a few years interval. Crop failure followed crop failure, from field to field yielded nothing but dead stalks.

The annual precipitation is estimated to be only ca 560 mm. The agriculture in Yerdona has often been referred to the showcase of Indian dry farming before the completion of Tungabhadra dam. The inmutability of the dry belt, extending vastly over the balk of interior Karnataka state along the Western Ghats, its vastness, the stark vastness of so much undifferentiated is owesome.

The hot eye of the sun evaporates the land's final drop of dew stored in the earth. The distinction between land and sky is always nothing but the straight-

Table 1-20 Trends in the prices of agricultural lands in Yerdona

Year	Paddy field(R.S.)	Land under light crops(R.S.)	Dry land (R.S.)
1957	300 ~ 400	250 ~ 300	100 ~ 150
1960	3,000 ~ 3,500	2,500 ~ 3,000	200 ~ 250
1970	5,000 ~ 6,000	5,000 ~ 6,000	400 ~ 500
1975	6,000 ~ 7,000	5,000 ~ 6,000	400 ~ 500
1980	8,000 ~ 10,000	8,000 ~ 9,000	$\geq$ 500

Data collected as of 11, Dec., 1980

lined horizon. With the onset of monsoon, it becomes, however, blurred with turbulent clowds in a dozen different quick, pitching and heaving squalls that shower and wash away fragile top soils of cultivated land. But the agricultural landscape and the way of life of villagers who clung to the bleak stretches of barren land have completely changed in many respects.

With the introduction of irrigation canal, agricultural activities of various kinds has taken a very decided turn for the better. Progress was visible everywhere in the village. The irrigation canal gave a new life to the villagers. One of the most remarkable trends is a growing tendency of population. There lived only 1,223 in 1951. With the opening of irrigation canal in 1957, Yerdona village and its environs gained in population being 1,540 in the census year of 1961. The labour intensive rice cultivation has created a vast amount of job opportunities, population has nearly doubled amounting to 3,173 in 1971. Population has thus jumped by 206% between the two decades. A throng of in-migrants swarmed to Yerdona and its environs. As of 1980, population size has been enumerated to be 5,354. There are several possible explanation for the radical increase of population still now under way at a phenomenal rate.

Firstly, the greater part of population increases could be attributable to social increase, viz, cumulative results of in-migrants who have settled down and established a brand new households in the village and its environs.

Secondly, social increase of population has rapidly thronged to Yerdona particularly when the dry farming was switched over to the labour intensive irrigation farming.

The band of first settlers as in *Kindi* near by Yerdona have played an innovative role in diffusing the know-how of rice cultivation in various ways widely over Yerdona and its environs. And as the farmers who once engaged in dry farming have gradually assimilated many new experiences and expertises on rice cultivation. Now that rice cultivation is the farming of labour intensive nature, so many job opportunities related to rice cultivation has been created.

Population influx puts increased pressure on a sometimes fragile rural environment especially, aggravation of sanitation, contaminated drinking water, disorderly expanding of residential blocks and cultural frictions has been keenly felt in recent years. Many of those over expanding communities in the village and even over to outskirts are generally segregated in their way of life, social communications.

Each community is mainly set up as community (*jati*)-wise. This is partly because they have their own way to make a living and are largely unprepared for major socio-cultural adjustments each other that has been strictly observed in their own way. Newly settled farmers are still more deprived, caught in the vicious circle of poverty. A few of them have succeeded in expanding holding size. They are mostly employed as an agricultural labourer, earning 3 to 6 rupees a day with which they can barely afford to make a from hand to mouth living.

Surging wave of immigration could well be termed as "Irrigation Rush". This has been caused by the construction of irrigation trunk line, viz, distributaries as No. 31/4 and No. 31/3-2 directly connected with Tungabhadra Left Bank Main Canal. Yields per acre for rice has been recorded at 3 quintals at the dawn of irrigation (1957–60) then, on the next stage, it jumped up to the higher level varying from 15 to 20 quintals of recent years, yields have levelled off at ca 25 quintals. The remarkable increases in productivity per acre have naturally been accompanied by the steady rise of price of agricultural land. This holds true specially of paddy field as clearly seen in Table 1–20, and there can be seen some remarkable upswings in the two decadal periods from 1960 to 1970.

But the most striking uptrend is the several years just after the opening of new era beginning in 1957. It is very interesting to note that the price of paddy field had dramatically increased, being from 25 to 26.7 times as high in 1980 as in 1957. Upturn trends also true of the land under light crops. But these trends are sharply contrasted by the counterpart for dry land, being 3.3 to 5 times as high in 1980 as in 1957.

Soaring prices of agricultural lands have more profound and possibly enduring impacts which are now occuring. The following impact must be emphasized here

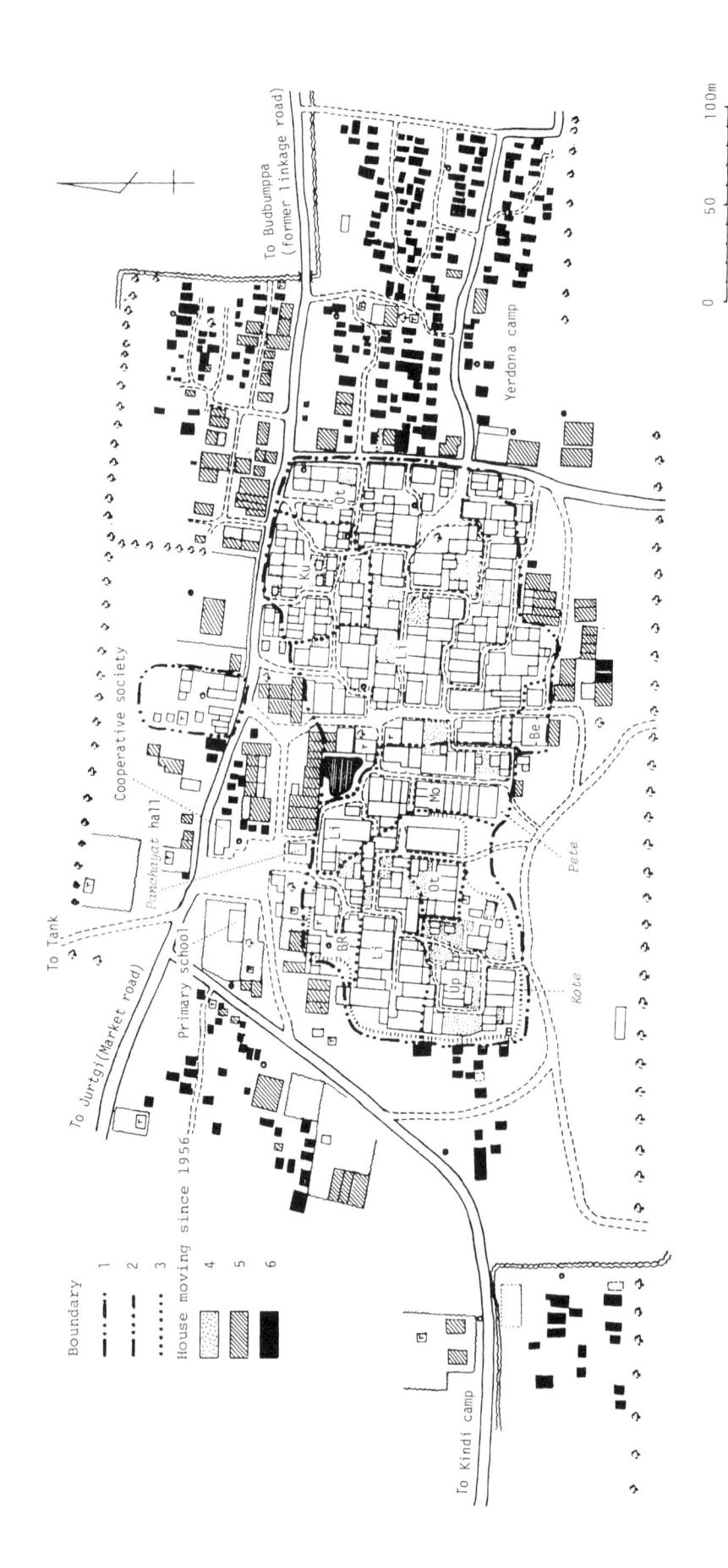

Fig. 1-19 Changing pattern of settlement in Yerdona

1. The former settement before 1955,
2. The original settlement,
3. Residential area by jati,
   Li: Lingayat, Ku: Kuruba, Be: Bedda,
   B.R.: Brahman and Rajiput, Or: Others,
   Mo: Moslem.
4. Blank space from which a house moved to the outside of the former settlement,
5. House moved from the former settlement,
6. House moved from other villages.

Table 1–21 Year settled of agricultural labourers in Yerdona and Kindi camps

Year settled	Yerdona camp		Kindi camp		Total	
1955						
1956						
1957	17	17			17	17
1958						
1959						
1960	7				7	
1961			1		1	
1962	5	18		9	5	27
1963						
1964	6		8		14	
1965	11				11	
1966	7				7	
1967	5	27		4	5	31
1968	4		4		8	
1969						
1970	5		3		8	
1971	2				2	
1972	4	18		5	4	23
1973	4		1		5	
1974	3		1		4	
1975	2		1		3	
1976	2	19	1		3	
1977	3		3	8	6	17
1978	1		3		4	
1979	6				6	
1980	5				5	
Total	99	99	26		125	125

Table 1–22 Spatial distribution of agricultural labourers settled in Yerdona camp

District	*Taluk*	Number of H.H.	
Raichur	Gangavati	5	
	Sindhnur	6	
	Kustagi	45	79
	Manvi	6	(46.7%)
	Koppal	12	
	Lingsugur	2	
	Raichur	3	
Bellery	Bellery	2	8
	Hospet	3	(4.7%)
	Harpanahalli	1	
	Siruguppa	2	
Bijapur	Hunugund	10	10(5.9%)
Bidar	Humnabad	2	2(1.2%)
Gulbarga	Chincholi	7	7(4.1%)
A.P.Hurnool	–	1	1(0.6%)
Local		62	62(36.7%)
Total		169	169(100.0)

on that the remarkable rise in land have made it difficult for the new comers in this village to purchase in additional land for the purpose of the expansion of holding size. Nobody wants to lease out his land especially paddy fields mainly because it produces a substantial yields enough to maintain a staple foods with even a few acres.

This explanation could be valid on referring to the column of holding size. In this process, soaring prices of land tends to differentiate middle holding classes into the higher and lower classes. In some cases, some of them have been reduced even to the coolie's class. It was not 1975 and afterward since a handful -big farmers have much concerned with a profit-oriented activities, placing major emphasis on rice cultivation combined with possibly the more profit yielding agribusiness and new line of business as milling, grain broker and bus owner. They could make a lot of money through an intensive cultivation of commercial crops as rice, groundnuts and often high yielding crops, together with non-agricultural income in a hasty way. Their business activities are very tactful in dealing with additional ventures and willing to pour enormous money into the construction of imposing big house with a finely decorated façade, new venture, banquet and other purposes sometimes in ostentatious ways. At a first glance, they seem to have climbed up social ladders with the strength of Samson.

On the other hand a mass of people live in a desolated state who have not benefited from a buoyant economy. An accelerating processes involved in rapid development and polarization of socio-economic class have naturally been accompanied by an increasing social costs this aspect is usually less known and liable to be underestimated. The problems are particularly magnified in those developing rural areas. The deprived peoples without a meagre land are composed of new comers and the landless *coolies* who have been reduced from lower middle on marginal farmer's class.

In fact the construction of Tungabhadra Dam Project has led not only to an increased job opportunities for the landless and deprived peoples, but accelerated polarization of the have and the have-not. At the same times, special attention must be paid to the another aspects of the phenomena that they are closely interwoven into the webb of rural life and seems to have supported an expanding economy. Symbiotic interdependences between the two sides have played a role in lifting up to the prosperous economy of present day. Either of them is indispensable for each other.

We will provide a concise review for one of the typical process of rural development based on the em-

Table 1-23 Spatial distribution of agricultural labourers settled in Kindi camp

Loc.No.	Adult M	Adult F	Children	Total	Former residence State/District	Former residence Village/*Taluk*	Year settled	Community (*Jati*)
1	2	1	3	6	Raichur	Bapur	1970	Muslim
2	1	1	–	2	Raichur	Boydaddi	1978	*Koppallu*
3	1	1	1	3	Raichur	Bapur	1964	*Byoda**
4	2	3	3	8	Raichur	Bapur	1964	*Byoda**
5	5	3	2	10	Raichur	Bapur	1960	*Byoda**
6	1	1	1	3	Raichur	Boydaddi	1978	*Lingayat*
7	1	1	2	4	Raichur	Bapur	1964	*Byoda**
8	1	2	3	6	Raichur	Bapur	1964	*Byoda**
9	–	5	3	8	Raichur	Bapur	1964	*Byoda**
10	1	1	1	3	Raichur	Bapur	1964	*Byoda**
11	1	2	1	4	Raichur	Bapur	1973	*Byoda**
12	1	3	3	7	A.P.	Doddisindhanoor	1977	–
13	1	2	–	3	A.P.	Hirichadu	1964	*Valmiki**
14	1	1	3	5	Raichur	Bapur	1964	*Valmiki**
15	2	2	–	4	Raichur	Vokreno	1970	*Kuruba*
16	1	2	2	5	A.P.	Godawal T.Q.	1976	*Naik*
17	1	1	–	2	A.P.	Godawal T.Q.	1968	*Byoda**
18	1	1	2	4	A.P.	Godawal T.Q.	1968	*Kuruba*
19	1	2	1	4	A.P.	Godawal T.Q.	1968	*Kuruba*
20	1	1	3	5	Raichur	Bengala	1977	*Lingayat*
21	2	1	2	5	Bellery	Huliyal	1975	*Valmiki**
22	1	1	3	5	Raichur	Bapur	1970	*Valmiki**
23	1	1	2	4	A.P.	Godawal T.Q.	1968	*Byoda**
24	2	2	3	7	Bellery	Huliyal	1974	*Madiwaru*
25	2	2	3	7	A.P.	Etlapura	1978	*Madiwaru*
26	1	3	3	6	Raichur	Kowdapur	1977	*Kobbe*
Total	35	46	50	131				

* Backward communities

Fig.1–20 Spatial distribution of agricultural labourers settled in Yerdona camp

48

pirical data. Particular attention is forcused on the interaction between supply and demand of labour forces.

With the advent of irrigation canal, many a job related to the expanded agricultural activities turned out to be in great need.

At the initial stage, Kindi camp had established as a pilot farm of rice cultivation. This settlement was the first evolving eventually into the full fledged "frontier line", from where all the farming practices, information and furthermore technical innovation have diffused widely and came into common use.

The first settlers coming from Andra Pradesh state called in agricultural labour from their original villages. This was partly because they have been engaged in rice cultivation and partly because few cultural frictions and conflicts expected to arise, as having same socio-cultural backgrounds in common. On the next stage, as the switching dry farming over to rice cultivation has smoothly been carried out, there was apparent the growing demand for additional agricultural labourers. Year-wise settlement of in-migrant in Yerdona and Kindi camp are shown in Table 1–21.

17 house holders came to Yerdona in 1957 with their dependents and nearly the same member of householders settled down from 1960 to 1964, and then the number jumped up at 27 from 1965 to 69, marking the highest peak in immigration history. Entering 1970's, there could be seen declining trends in numbers. But, putting the figures of Yerdona and Kindi camps together, there has been a constant stream of in-migration with some upheaval from 1965 to 1969. Spatial distribution of in-migrants settled in Yerdona camp and Kindi camp with some details of social attributes are indicated in Table 1–22, Table 1–23, and Fig. 1–20.

In-migrants settled in Yerdona camp have come from mainly Raichur, Bellery and its environs in Karnataka, amounting to ca 50% of the total. As for the case of *Kindi* camp, they have streamed out of the dry belt of north Karnataka, Raichur and Andhra Pradesh en masse. Most of them belong to so called backward community as *byoda* and *valmiki*.

Their camp, mainly composing of shabby hut thatched with "*appu*", a leaves of wild grasses grown in the marsh land and swamp is dotted on the bank along the irrigation distributary. Pulling factor just as have discussed, combined with pashing factor as severe drought, has accelerated the growing influx of agricultural labourers.

The need for hired labourer could not be met even by onward surges of the deprived peoples at the peak time. In response to excess demands for labour supply, work forces have dramatically been swollen from year to year. The job seeker, lured by a buoyant story about the irrigation booms are surging into the Yerdona, once extremely poverty stricken settlement.

Another bands of labourers come in a steady stream of battered carts en masse piled high with all their belongings and goods, whipping with lashes the bullocks at a gallop on the dusty road. They are seasonal worker afloating from place to place to be hired.

They usually temporarily settle down near by the cross road thronged with villagers, where they have easy access to dependable informations of various kinds, water and open markets. A band of people bed down and curl up on the patch of grass mat in the open fields in a land that is singularly hard to endure. Those casual labourers are going to meet the excess demands of working forces on season.

Ever-expanding economy provides jobless peoples with a sum of livelihood. And a handful farmers of the strong willed and gutsy sort create a demand for labour that outstrips even the current flood of job seekers. "Job rush" coincides with economic booms caused by irrigation rush. Demand and supply of labour forces are going on hand in hand.

Irrigation and its economic consequences account for prosperity and ever-expanding agriculture that has sprung up from the immutability of bleak dry farming, a striking factor in its self so far witnessed of sharply constrasted scene between poverty-stricken small farms and large commercial farmers.

The precious experiences make us more confident about the future, if the sustained efforts to improve changing environmental settings and further develop rural regions just as in Yerdona would result in offering some crucial solution to many of the most pressing dilemma in Indian economy.

In those context we take optimistic stance despite the huge array of problems, now confronting rural India. For the time being, it remains to be seen whether the experiences of Yerdona would turn out to be dependable answer to the balanced growth of rural areas.

References

Command Area Development Authority, Tungabhadra Project (1976): *Two decades of development, Tungabhadra Project.* 49p.

Command Area Development Authority, Tugabhadra Project (1976): *Report on re-examination of cropping pattern under TBP, Left Bank Canal.* 116p.

Krishna Water Disputes Tribunal (1973): *The report of the Krishna water disputes with the decision.* Vol. 1, 123p.

Singh, Tapeshwar (1978): *Drought prone areas in India.* New Delhi, People's publishing house, 124p.

CHAPTER 2

POPULATION AND OCCUPATIONAL STRUCTURE OF MALNAD VILLAGES

1. Population and Occupational Structure

Kurubathur and Yadavarahalli were selected for the sample survey studies as a typical *Malnad* villages in a paddy and plantation crop growing area. Kurubathur is 19 km from Sakleshpur, *Taluk* headquarters, and 58 km from the District headquarters in Hassan. Kurubathur could be identified as a two-row street village, which is equivalent to Strassendorf in German terminology.

This village is not only the key settlement, but also plays integral part of much importance in the socio-economy of Kurubathur Group *Panchayat* (K.G.P.) consisting of nine villages, namely Yadavarahalli, Bellur, Nidigere, Ummathur, Echalahalli, Hadalahalli, Bommanakere and Vodarahalli. Before 1940 there was one *panchayat* per one village as a rule. But after 1940 Kurubathur was selected as the main village of Group *Panchayat* on the central place. The Group *Panchayat* system was reorganized in 1962.

But its location has remained unchanged. The major reasons for grouping 9 villages into one *panchayat* were accessibility and economic gravity to this village.

The geographical area of Karubathur and Yadavarahalli are 544.07 acres and 561.00 acres respectively (Table 2–1). Total area of the nine villages forming the K.G.P. adds up to 7,267.31 acres. The total population of K.G.P. was 2,026 in 1951, 2,744 in 1961, 3,359 in 1971 and 3,623 in 1981. The total population in 1981 amounts to 1.78 times greater than that in 1951 on the whole. The population in Kurubathur amounts to 416 and 290 in Yadavarahalli.

The population is divided into two categories, ie, worker and non-worker. In both 1961 and 1971 census, all the workers were classified into nine industrial categories according to their economic activity, but marginal workers were added as a separate category in the 1981 census. Table 2–2 shows the number and percentage distribution of the three census for each village. It reveals clearly the facts as follows;

1) Of the nine villages, economic activity of Kurubathur is much more varied in its functions than any of the other villages. Consequently, the occupational set-up of Kurubathur can be described as multi-functional and thus the weight of primary activities (e.g. categories I and II) are of considerably less importance compared with the other villages.

2) As for Yadavarahalli, its functional categories are less diversified in their variety. In the 1961 census, over 80% of workers were engaged in primary activity. However, there has been a tendency for the number of cultivators to decrease whilst the number of agricultural

Table 2–1 Increase of population by village in Kurubathur Group *Panchayat*

Village	Geographcal area(acre)	Population 1951	1961	1971	1981	Population increase in 1951-1981(%)
Kurubathur	544.07	277	387	398	416	150.2
Yadavarahalli	561.00	149	201	241	290	161.7
Bellur	947.93	364	464	460	460	126.4
Nidigere	1,024.19	300	410	422	470	156.7
Ummathur	1,420.30	159	330	386	381	239.6
Echalapura	404.08	86	131	150	152	176.7
Hadarahalli	1,033.31	423	403	907	1,001	236.6
Bommanakere	879.05	240	389	356	413	172.1
Voddarahalli	453.38	28	29	39	40	142.9
Total	7,267.31	2,026	2,744	3,359	3,623	
Population increase from 1951(%)		(100.0)	(135.4)	(165.8)	(178.8)	

Table 2–2 Population and occupational structure by village in Kurbathur Group *Panchayat*

		\multicolumn{6}{c}{Kurubathur}	\multicolumn{6}{c}{Yadavarahalli}
		\multicolumn{2}{c}{1961}	\multicolumn{2}{c}{1971}	\multicolumn{2}{c}{1981}	\multicolumn{2}{c}{1961}	\multicolumn{2}{c}{1971}	\multicolumn{2}{c}{1981}
Households		88	100.0	71	80.7	78	88.6	35	100.0	40	114.3	53	151.4
Total Population		387	100.0	398	102.8	416	107.5	201	100.0	241	119.9	290	114.3
Male		207		222		203		108		117		143	
Female		180		176		213		93		124		147	
Worker*	I	34	(22.4)	20	(14.5)	26	(16.0)	75	(76.5)	42	(56.8)	47	(38.8)
	II	4	(2.6)	41	(29.7)	74	(45.7)	4	(4.1)	27	(36.5)	72	(59.5)
	III	17	(11.2)	4	(2.9)			8	(8.2)	3	(4.1)		
	IV	21	(13.8)	–	(–)	3**	(1.9)	1	(1.0)	–	(–)		
	V	5	(3.3)	8	(5.8)			2	(2.0)	–	(–)		
	VI	25	(16.4)	24	(17.4)	59***	(36.4)	1	(1.0)	–		2***	(1.6)
	VII	23	(15.1)	27	(19.6)			4	(4.1)	–	(–)		
	VIII	–	(–)	3	(2.1)			–	(–)	1	(1.3)		
	IX	23	(15.1)	11	(8.0)			3	(3.1)	1	(1.3)		
Total		152	(100.0)	138	(100.0)	162	(100.0)	98	(100.0)	74	(100.0)	121	(100.0)
Marginal Workers						–						–	
Non-Workers		235		260		254		103		167		169	

		\multicolumn{6}{c}{Echalapura}	\multicolumn{6}{c}{Hadalahalli}
		\multicolumn{2}{c}{1961}	\multicolumn{2}{c}{1971}	\multicolumn{2}{c}{1981}	\multicolumn{2}{c}{1961}	\multicolumn{2}{c}{1971}	\multicolumn{2}{c}{1981}
Households		21	100.0	23	109.5	23	109.5	115	100.0	142	123.5	195	169.6
Total Population		131	100.0	150	114.5	152	116.0	661	100.0	907	137.2	1,001	151.5
Male		62		73		71		362		473		494	
Female		69		77		81		299		434		507	
Worker*	I	25	(75.8)	35	(70.0)	26	(45.6)	148	(75.5)	227	(69.8)	172	(50.4)
	II	4	(12.1)	12	(24.0)	8	(49.1)	3	(1.5)	5	(1.5)	149	(43.7)
	III	–	(–)	–	(–)			28	(14.3)	42	(12.9)		
	IV	3	(9.1)	–	(–)	2**	(3.5)	6	(3.1)	–	(–)	4**	(1.2)
	V	–	(–)	1	(2.0)			–	(–)	9	(2.8)		
	VI	–	(–)	–	(–)	1***	(1.8)	–	(–)	9	(2.8)	16***	(4.7)
	VII	–	(–)	–	(–)			–	(–)	8	(2.5)		
	VIII	–	(–)	–	(–)			–	(–)	5	(1.5)		
	IX	1	(3.0)	2	(4.0)			11	(5.6)	20	(6.1)		
Total		33	(100.0)	50	(100.0)	57	(100.0)	196	(100.0)	326	(100.0)	341	(100.0)
Marginal Workers						28						212	
Non-Workers		98		100		67		465		581		448	

* The worker are classified into nine functional categories as follows:
  I; Cultivator, II; Agricultural labourer, III; Mining, quarring, livestock, forestry, hunting, plantation; and allied activities, IV; Household industry,
  V; Manufacture other than household industries, VI; Construction, VII; Trade and commerce, VIII; Transport, storage, communication,
  IX; Other services
** Va; Household industry, manufacturing, processing, servicing and repairs
*** III, IV, Vb and VI to XI, Other workers

Chap.2 POPULATION AND OCCUPATIONAL STRUCTURE OF MALNAD VILLAGES

	Bellur						Nidigere						Ummathur					
	1961		1971		1981		1961		1971		1981		1961		1971		1981	
	86	100.0	77	89.5	84	97.6	92	100.0	77	83.7	102	110.9	59	100.0	82	139.0	74	125.4
	464	100.0	460	99.1	460	99.1	410	100.0	422	102.9	470	114.6	330	100.0	386	117.0	381	115.5
	247		239		235		209		189		219		155		208		197	
	217		221		225		201		233		251		175		178		184	
	173	(72.4)	53	(28.6)	80	(37.0)	171	(75.0)	82	(40.8)	31	(16.8)	38	(33.3)	41	(21.5)	67	(39.0)
	5	(2.1)	86	(46.5)	121	(56.0)	22	(9.6)	107	(53.2)	142	(76.8)	46	(40.4)	12	(6.3)	99	(57.6)
	54	(22.6)	24	(13.0)			10	(4.4)	–	(–)			23	(20.2)	73	(38.4)		
	–	(–)	–	(–)	4**	(1.9)	4	(1.8)	–	(–)			2	(1.8)	21	(11.1)		
	–	(–)	6	(3.2)			6	(2.6)	2	(1.0)	–		–	(–)	26	(13.7)		
	–	(–)	8	(4.3)	11***	(5.1)	–	(–)	2	(1.0)	12***	(6.4)	4	(3.5)	1	(0.5)	6***	(3.4)
	–	(–)	1	(0.5)			–	(–)	–	(–)			–	(–)	5	(2.6)		
	–	(–)	1	(0.5)			2	(0.9)	2	(1.0)			1	(0.8)	–	(–)		
	7	(2.9)	6	(3.2)			13	(5.7)	6	(3.0)			–	(–)	1	(0.5)		
	239	(100.0)	185	(100.0)	216	(100.0)	228	(100.0)	201	(100.0)	185	(100.0)	114	(100.0)	190	(100.0)	172	(100.0)
					–												228	
	225		275		244		182		221		285		216		196		107	

	Bommanakere						Voddarahalli						Total		Percentage	
	1961		1971		1981		1961		1971		1981		(1981)			
	66	100.0	65	98.5	79	119.7	5	100.0	6	120.0	5	100.0	693			
	389	100.0	356	91.5	413	106.2	29	100.0	39	134.5	40	137.9	3,623		(100.00)	
	211		175		212		15		16		18		1,792			
	178		181		201		14		23		22		1,831			
	72	(45.0)	41	(31.3)	127	(95.5)	6	(50.0)	3	(18.8)	8	(100.0)	584	41.87	16.12
	–	(–)	69	(52.7)	–	(–)	–	(–)	13	(81.2)	–	(–)	685	49.10	18.91
	11	(6.9)	5	(3.8)			1	(8.3)	–	(–)	–	(–)			
	1	(0.6)	–	(–)			–	(–)	–	(–)	–	(–)	13**	0.93	0.35
	1	(0.6)	4	(3.0)			–	(–)	–	(–)	–	(–)			
	1	(0.6)	9	(6.9)	6***	(5.5)	–	(–)	–	(–)	–	(–)	113***	8.10	3.12
	4	(2.5)	–	(–)			–	(–)	–	(–)	–	(–)			
	–	(–)	–	(–)			–	(–)	–	(–)	–	(–)			
	70	(43.8)	3	(2.3)			5	(41.7)	–	(–)	–	(–)			
	160	(100.0)	131	(100.0)	133		12	(100.0)	16	(100.0)	8	(100.0)	1,395	(100.0)	38.50
					119						10		371		10.24
	229		225		161		17		23		22		1,857		51.26

53

Table 2–3 Hierarchical structure of socio-economic functions in the nine villages of Kurubathur Grorp *Panchayat*

	Kuruba-thur	Bellur	Nidigere	Ummathur	Echala-pura	Hadrahalli	Yadavara-halli	Bommana-kere	Vadara-halli	Total	The year established
Business tax (Rs)	735.0		15.0			130.00		30.00		910	
Assessment per capita (Rs)	7.20	6.67	6.51	4.58	4.38	4.18	3.79	3.47	2.56	5.59	
Assessment per household (Rs)	38.42	36.50	29.99	23.61	28.93	21.48	19.82	19.00	20.45	26.49	
Drinking water	W	W	W	W	T	W	W	W	W		1973
Road	P	P	P	P	K	P	P	P	P		1963, 1976
No.of household electrified	22	9	8	16		26	6			87	1972
Education — Primary school & No.of pupil.	○ 100	○ 58	○ 62	18	12	94	23	○ 102	12	481	1942
Middle school & No.of pupil.	○ 113	14	14	12	8	18	16	18	6	219	1975
High school & No.of student.	○ 40	6	3		4	3	6	9		71	1979
Nursery school & No.of children.	51							13		64	
Woman society	WS							WS			
Youth society	YS					YS					
Post office	PO		PO								1930
Bank	B										1980
Primary health unit and clinic centers	PHU		CC					CC			1920
Telephone exchange	T										1978
Veterinary dispensary	VT										1981
Cooperative society	CS										1951
Tourist bangalow	TB										
Group panchayat office	GPO										1940 reorganized in 1962
No.of scheduled castes	48	74	229	132	42	206	79	121		931	
No.of scheduled tribes	23	19		9			70		8	129	
No.of *Janata* house constructed	9					18	9	10		46	
No.of marginal workers				2	28	212		119	10	371	
No.of agricultural labourer	74	121	142	99	28	149	72			685	
Gross area planted under	80.10	69.19	54.30	227.27	29.13	140.22	106.01	281.36	94.37	1,084.33	
Coffee in acre (%)	(7.4)	(6.4)	(5.0)	(21.0)	(2.7)	(12.9)	(9.8)	(26.0)	(8.7)	(100.0)	

○ mark shows location of a school.

labourers called *coolie* have increased from 4 in 1961, to 27 in 1971 and to 72 in 1981. This phenomenon could be explained by the facts that greater part of extent of land has been concentrated in the hands of the fewer land owners who have ventured into speculative plantation economy of coffee and cardamom in recent years. The revenue from profitable plantation economy has been ploughed back into acquiring more agricultural land and hiring more labourers.

3) Economic activity in the four villages, Bellur, Nidigere, Ummathur and Hadarahalli are less multifunctional in its variety compared with the main village of Kurubathur. Their non-primary function has been gradually absorbed by the primate settlement, Kurubathur. Thus, economic activity in these villages tends to be mono-functional.

4) Other villages such as Echalapura and Voddarahalli can be termed as mono-functional settlement. In Voddarahalli all the workers specialize completely in primary occupation. 81.2% of the workers are classified as agricultural labourers.

Table 2–3 indicates hierarchial structure of the socio-economic functions in the nine villages. The primary indices available showing economic strength of each village are business tax and several assessment data which include library cess, education cess and health cess. Business tax suggests of economic potentials of the villages concerned. Kurubathur comes at the top of the list with the percentage of 80.77 (Rs 735). The amount of business tax levied in the other villages (14.29% Rs 130) is far lower than in Kurubathur. This shows predominant economic influences of Kurubathur over the other villages. Bommanakere and Nidigere have only negligible share of 3.30%, 1.65% respectively. These three villages are auxiliary commercial centers, supporting the primate key settlement of Kurubathur. Socio-economic strength of each village is well reflected in assessments per capita and household. The strength of each village defined as the amount of assessment levied is arranged according to the size of assessment per capita. Assessment per household parallels fairly with that of per capita.

The economic order of the nine villages is closely related to the distribution of educational facilities, socio-economic institutions and societies. Most of public facilities are mainly concentrated in Kurubathur Village. The auxiliary sub-center are Bellur, Nidigere, Hadrahalli and Bommakere.

Kurubathur has almost monopolized various basic functions and among various functions, Group *Panchayat* Office, Post Office, bank, Primary Health Unit and Veternary Dispensary are of primate importance. These facilities meet the basic needs to neighbouring villages and play an important part in integrating and unifying each village.

Chap.2 POPULATION AND OCCUPATIONAL STRUCTURE OF MALNAD VILLAGES

Table 2–4 Households classified by religion and community in Kurubathur, 1960

Religion	Community	No. of Households	Population Total	Male	Female
Hindu	*Lingayat*	11	58	34	24
	A dikarnataka	6	24	14	10
	Vokkaliga	5	31	21	10
	Brahmin	2	7	5	2
	Banajiga	2	12	9	3
	Kshatriya	2	3	–	3
	Nayar	2	12	5	7
	Maharashtra	2	14	6	8
	Jyothipana	1	3	1	2
	Besta	1	8	2	6
	Viswakarma	1	8	4	4
	Nainda	1	3	2	1
	Kuruba	1	1	1	–
	Total	37	184	104	80
Muslim	*Moplah*	24	125	60	65
	Syed	5	27	15	12
	Sunny	4	23	10	13
	Mughal	1	5	1	4
	Labbe	1	3	2	1
	Total	35	183	88	95
	Grand total	72	367	192	175

Availabilities of related basic functions in the key settlement, Kurubathur attracts many people not only from the nine villages belonging to Group *Panchayat* but also from an extensive neighbouring region. Of public facilities, Primary Health Unit is particularly important. This unit dates back to the year of 1929 when locally founded dispensary was opened. Medical service area of P.H.U. extends over 21 main villages and 27 sub-villages, with clinic centres located in Nidigere and Bommanakere. There are eleven members in the unit; 1 doctor (assistant medical health officer), 2 compounders (pharmacists), 4 health workers and 4 paramedical workers (class fourth). Daily patients average 20 to 25 persons, adding up to 2,000 to 3,000 patients in a year. P.H.U. provides necessary care for sufferers of such diseases and illness as anaemia malnutrution and maleria, diarrhoea, dysentery asthma and so forth.

The distribution of agricultural labourers and marginal workers in the nine villages is closely related to the presence of estate economy specializing mainly in the production of coffee and cardamom.

The estate economy is heavily dependent on a massive, permanent agricultural and seasonal labour forces. Most of them belong to the so called 'weaker section', for whom 46 *Janatha* houses were constructed.

Differential population growth between the villages could be partly explained by the steady expansion of plantation economy. Marked increase in the number of agricultural labourers is responsible for the social increase of population. Another pull-factor of population is attributable to thriving commercial activities which has been associated with expansive estate economy.

Examining village-wise population growth between 1951 and 1981, it comes to clear that Ummathur and Hadarahalli have recorded a remarkable growth; their population being nearly 2.4 times larger in 1981 than that in 1951 (Table 2–1). The gross areas planted under coffer in Ummathur and Hadarahalli stand second and third largest among the nine villages. The striking increase in population in these villages should be attributed to the expansion of the plantation economy. Population growth clearly reflects the gross area of coffee grown in each village.

2. Distribution of Household by *Jati*

There were 72 households and 367 in total population in Karabathur in the 1961 census. There were 72 families; 37 were Hindus and 35 Muslims (Table 2–4). Of the 13 Hindus and 5 Muslims, the *Lingayat* and *Moplah* communities were the largest in term of the size of household and population, which is followed by the *Vokkaliga, Adikarnataka,* and *Syed* communities. In this respect Kurubathur could be defined as a multi-*jatis* village. However, among the various *jatis Lingayat, Vokkaliga, Adikarnataka* and *Moplah* occupy predominant share.

Taking the distribution of landholding by *jati* into consideration, supreme dominance of Lingayat over the remaining *jati*s can be evidently seen in Table 2–5. In 1960 Lingayat has 261 acres of land. This is 77.1% to the total extent of land in Kurubathur. The total population of the Lingayat community comprised only one-third.

The results of our survey in 1982, tabulated in Ta-

55

Table 2-5 Distribution of landholding by main community in Kurubathur, 1960

Community	Below 1 acre	1.0-2.4	2.5-4.9	5.0-7.4	7.5-9.0	10.0-14.9	More than 15	Total extent and percentage of land (%)		Total number and percentage of household (%)	
Lingayat	–	4.11 (2)	4.00 (1)	5.22 (1)		10.0 (1)	238.31 (4)	261.64	77.1	9	36.0
Muslim	–	7.95 (4)	3.00 (1)		32.41 (4)	14.85 (1)	–	58.21	17.1	10	40.0
Vokkaliga	–	–	2.90 (1)	6.35 (1)	8.40 (1)	–	–	17.65	5.2	3	12.0
Adikarnataka	1.00 (2)	1.00 (1)	–	–	–	–	–	2.00	0.6	3	12.0
Total	1.00 (2)	13.06 (7)	9.90 (3)	11.57 (2)	40.81 (5)	24.85 (2)	238.31 (4)	339.50	100.0	25	100.0

Note : Figures in brankets is the number of household.

Table 2-6 Landholding size by community in Kurubathur and Yadavarahalli, 1982

Community	0 K	0 Y	0.1-0.9 K	0.1-0.9 Y	1.0-2.4 K	1.0-2.4 Y	2.5-4.9 K	2.5-4.9 Y	5.0-7.4 K	5.0-7.4 Y	7.5-9.9 K	7.5-9.9 Y	10.0-14.9 K	10.0-14.9 Y	15.0-19.9 K	15.0-19.9 Y	20.0-24.9 K	20.0-24.9 Y	25.0-49.9 K	25.0-49.9 Y	More than 50.0 K	More than 50.0 Y	Total K	Total Y	K+Y
Lingayat	1	1			4	2	1	3			2	2	2	1	1	2			3	2	2	1	15	14	29
Vokkaliga	9						3	2	1	1		1			1		1				1		13	7	20
Muslim	1				1				1					1									20		20
Bestru						1			1		1													3	3
Vaishanava													1										1		1
Setty	4								1														1		1
ViswaKarma					1				1														2		2
Bogies										1														1	1
Gangamantha		1		1		2		4																8	8
Kumbara						1		1											2						
S.C.	1	12					2																3	12	15
Adikarnataka	2				1		1																4		4
Besteir								1																1	
Besta					1																		1		1
Bunt						1																	1		1
Madivalas					1																		1		1
Achar	2		1																				3		3
Billava	2	1																					2	1	3
Nayar	2																						2		2
Brahmin	2																						2		2
Bandari	1																						1		1
B.T.		1										1												2	2
Unknown	2						1																3		3
Total	45	16	1	1	10	6	8	11	3	3		2	3	2	2	3		1	3	2	3	1	81	51	132
K + Y	61		2		16		19		6		8		5		5		1		5		4		132		

Note : K; Kurubathur, Y; Yadavarahalli

Table 2-7 Changes of landholding size in Kurubathur, 1961 to 1982

Landholding Size	1961 No. of households	1961 Percentage	1982 No. of households	1982 Percentage
Landless	16	22.2	45	55.6
Less than 2.5	16	22.2	11	13.6
2.5 – 5.0	12	16.7	8	9.9
5.0 – 10.0	16	22.2	6	7.4
10.0 – 15.0	6	16.0	3	13.6
More than 15.0	6		8	
No. of sampling	72	100.0	81	100.0

ble 2–6, clearly indicates that Lingayat has the biggest share of agricultural land among the various *jati*s. The situation was the same in the case of Yadavarahalli. Most of big farmers with more than 30.0 acres of land are estate owners.

They manage plantation estates where the major crops are now highly specialized in cardamom and coffee. The estate owners who held more than 50.0 acres in 1982 primarily belonged to the *Lingayat* and *Vokkaliga*.

A closer examination of the distribution of classwise landholding in Kurubathur since 1961 reveals that a greater part of agricultural land has been concentrated in the hands of a few big farmers (Table 2–7). There has also been a clear increase in the size of landless class from 16 (22.2%) to 45 (55.0%). To a lesser degree a substantial decrease can be seen in the marginal and medium classes, while the number of big farmers having more than 15 acres remained almost unchanged. The recent trends in the distribution of landholding could be largely explained by the fact that most of big farmers who had been engaged in plantation economy were exempted from the enforcement of Karnataka Land Reform Act.

Being immune from the so-called ceiling on landholding prescribed in the act, they could afford to plough back part of their profits into the aquisition of additional land. Thus the polarized distribution of landholding is mainly attributed to this clause which has, in effect, greatly favoured the estate owners. A few *jati*s as Lingayat, Vokkaliga and some of Muslims have steadily gained significant position in the village economy.

CHAPTER 3
RECENT TRENDS IN RICE CULTIVATION AND PLANTATION ECONOMY OF THE WESTERN GHATS

1. Macroscopic Overview

The distribution of village-wise land categories in the Kurubathur Group *Panchayat* suggest a cogent clue to follow up the salient features of ecological balance between man and natural environment. Farming system as a man-made ecosystem seems to be closely related with the predominant land use patterns in this region.

The distribution of village-wise land categories are tabulated in Table 3–1. It could be reduced that Kurubathur Group *Panchayat* shows the land use patterns shown as D^2-w^1-Co^1-Gr^0. This reads two different ways. The first point is the fact that dry land and grazing land are substantially meant to be the same and that the intensive land categories as wet land and coffee comprise nearly one-third of the total acreages. Unirrigated barren land and poorly grown grass land for livestock grazing occupy a considerable extent of land (41.8%). When waste land is added to it, it comes up to 46.5%. Nearly half the total area is of little account from the view point of land use intensity.

These categories of lands, however, are the resources for various agricultural activities. Seemingly barren lands provide a shrubery for fuel, green manures fertilizing paddy fields. Undulating verdure of hillsides has always gladdened weary eyes of villages and prevent the surface from soil erosion caused by torrential heavy rainfalls during the monsoon season.

But most important use of these lands have proved to be grazing land and unirrigated crop land. Density of bovine varies from 1.03 to 0.34 per acre of the geo-

Table 3–1 Distribution of village-wise land categories and land use patterns in Kurubathur Group *Panchayat*

Villages	Geogr. area	Wet land	Garden	Coffee	Dry land	Grazing land	Waste land	Land use pattern *	Bovine No.	Density	Acre per population Geogr. area	Wet land
Kurubathur	544.07 (100.0)	97.14 (17.85)	0.06 (0.06)	80.10 (14.72)	185.16 (34.00)	119.36 (21.94)	4.26 (0.78)	D^3-Gr^2-W^1-Co^1	413	0.76	1.31	0.23
Yadavarahalli	56.00 (100.0)	148.37 (26.45)	– (–)	106.01 (18.90)	106.07 (18.91)	140.16 (24.98)	2.36 (0.40)	W^2-Gr^2-D^1-Co^1	384	0.68	1.93	0.51
Ummathur	1,420.3 (100.0)	255.22 (17.97)	– (–)	227.27 (16.00)	411.11 (28.95)	198.01 (13.94)	134.22 (9.50)	D^2-W^1-Co^1-Gr^1	479	0.34	3.73	0.70
Bellur	947.13 (100.0)	101.39 (10.70)	– (–)	69.19 (7.31)	90.39 (9.54)	304.07 (35.91)	35.15 (3.71)	Gr^3-W^1-D^0-Co^0	913	0.96	2.06	0.22
Hadrahalli	1,032.31 (100.0)	175.14 (16.97)	0.24 (0.02)	140.22 (13.58)	254.21 (24.63)	62.37 (6.04)	81.08 (7.85)	D^2-W^1-Co^1-Wa^0	1,061	1.03	1.03	0.17
Vadarahalli	453.38 (100.0)	90.23 (19.09)	– (–)	94.37 (20.81)	195.00 (43.01)	20.08 (4.43)	27.12 (5.98)	D^4-Co^2-W^1-Wa^0	113	0.25	11.34	0.44
Bommanakere	879.05 (100.0)	118.28 (13.46)	1.24 (0.14)	281.38 (32.00)	285.15 (32.44)	71.15 (8.09)	9.16 (1.04)	D^3-Co^3-W^1-Gr^0	474	0.54	2.13	0.29
Nidigere	1,024.19 (100.0)	122.07 (11.92)	5.15 (0.50)	54.30 (5.30)	323.15 (31.55)	95.21 (9.30)	45.19 (4.41)	D^3-W^1-Gr^0-Wa^0	350	0.34	2.18	0.26
Echalapura	404.08 (100.0)	87.04 (21.54)	– (–)	29.13 (7.21)	174.38 (43.15)	62.16 (15.38)	– (–)	D^4-W^2-Gr^1-Co^0	194	0.48	2.66	0.57
Total	7,267.91 (100.0)	1,197.28 (16.47)	7.29 (1.00)	1,084.33 (14.92)	1,927.62 (26.52)	1,110.97 (15.28)	340.28 (4.68)	D^3-W^1-Gr^1-Co^1	4,981	0.60	2.01	0.33
Land revenue per acre (Rs)		2.23	5.55	14.86	7.47	0.72						

* Capital letter indicates land categories as follow:
  W; Wet, G; Garden, Co; Coffee, D; Dry land, Gr; Grazing land, Wa; Waste land
  Suffix indicates percentage distribution as follow:
  0; 0-9.9%, 1; 10-19.9 %, 2; 20-29.9 %, 3; 30-39.9 %, 4; 40-49.9 %

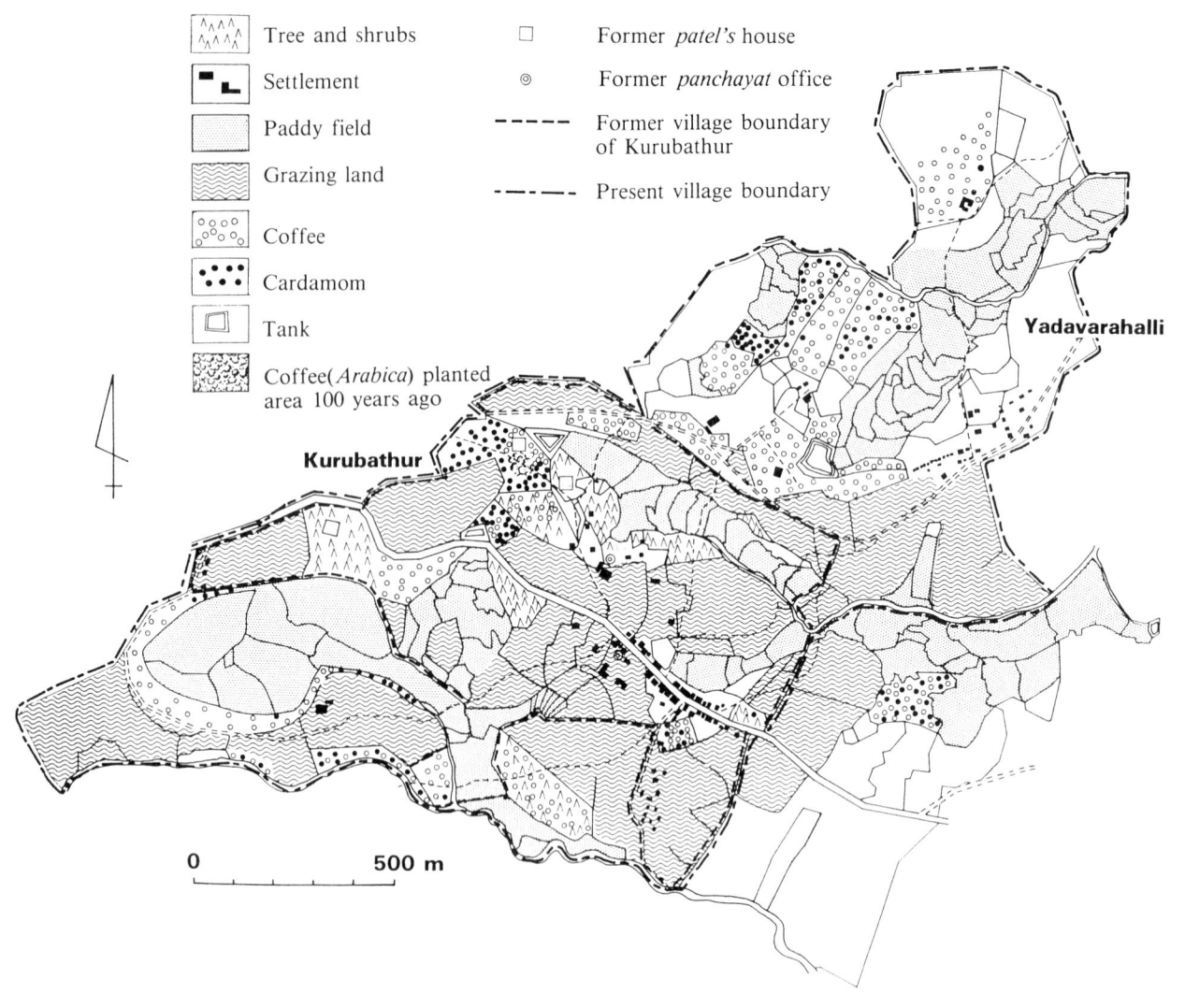

Fig. 3–1 Land use in Kurubathur and Yadavarahalli

graphical area. The remaining cultivable wet land is limited in extent, all the more reason for being intensively utilized. Therefore it was reasonable choice that coffee and cardamom was introduced as an intensive commercial crops.

Geographical area per capita varies widely from 11.34 to 1.03 acres with the average of 2.01 acres. Wet land per capita average 0.33 acre. These figures indicate the carrying capacity of land. As a whole, various land use patterns could be reduced as D^2-W^1-Co^1-Gr^0.

2. Recent Trends in Agricultural Land Use

The two sample villages, Kurubathur and Yadavarahalli are situated on the undulating hills with the altitude to 950m, consisting of widespread erosional surface. Recent annual rainfalls average about 3,300mm (1979–1981). The South-west Monsoon usually bring about heavy rains more than 90% of annual rainfalls. These two villages receive a few drops of rains from December to next April except the pre-monsoon and the post-monsoon seasons. Seeing from these unevenly distributed rainfalls, being concentrated in South-western monsoon period, the region covering sample surveyed villages falls into the humid savanna climate.

The topographical landscape could be characterized by the several undulating hills running from northwest to south-eastern direction and valley plains. To the south, the river flow down the Hemavathi River, branched off from the Cauvery River. The slope of the hills can be divided into two parts; the upper gentle slope with gradient of 1°–10° and the basal steep slope with gradient of about 30°. The upper slopes of hills have been utilized as grazing land which include non-irrigated (dry) land, waste and a part of afforested areas (Fig. 3–1).

Among the natural vegetations covering the upper parts of slopes, green grass is most important for summer grazing of bovines. However, natural growth of grasses is in poor condition. They stand at 3 to 5 cm in height. This is partly responsible for over-grazing in recent years.

As the basal parts of steeper slopes are well drained, there grows coffee, cardamom and other tropical plantation crops as banana, ginger. For the healthy and vigorous growth of coffee and cardamom plant, they need shade trees protecting them from strong winds and sunlight. A quick-growing trees such as *Erhthrina* are planted for temporary shading of cardamom plants and others tall spreading trees for permanent shade. *Dadap* is commonly used as a lower-canopy shade tree for coffee plants.

A full-grown coffee tree itself provides permanent shade for cardamom. So the cardamoms are grown under the coffee plants and are usually grown on the lowest part of slopes. After a few years of productive life, the cultivated areas under cardamom are often abandoned in favor of more suitable area for an increased crops. This is based on shifting cultivation. Ginger is grown as a side crop of coffee.

On the other hand, paddy grows on the flat and widespread valley plains. The uppermost parts of valley plains are dammed up by the earthen banks to store a water streaming down from the valley heads. As shown on the map, there are 10 tanks corresponding to each valley plains. Each valley plain has a tank for irrigation for its own purposes.

Thanks to these tank farmers can pool unevenly distributed annual rainfalls and make best use of it as much supply of water as they need all the year around. Most of farmers usually raise one crop of paddy. Double cropping plots totals only 2.

The changing land use patterns seem to be closely related with the spatial distributions of settlement in Kurubathur. Special attentions should be paid to the following point;

1) In former time, the settlements were located on the lower part of slopes. This is due to easy access to water for daily uses and to paddy fields as well. The most typical cases among these settlement locations can be seen in Kanarahalli estate. The four farmhouses were located just amid paddy fields. While the former *patel's* house, the first settler of this village was situated at the lower part of the slope facing water tank. For the householder this location was the shortest and most easiest shortcut to water and paddy fields he owned.

In addition, he could take good care of coffee arabica with the extent 2 acres and 15 gentes. All the fully-fledged plantation estates in the sample villages originated in this cradle land of coffee. These are among the most important geographical factors in determining location of settlement at that times.

2) But in those three decades, settlement pattern has been completely changed by ever-expanding plantation economy. In Kanarahalli, the four houses has given place to the newly constructed farm house (toft) on the edge of the hills.

On the whole, the scattered settlement has been concentrated on the main road taking form into two-row linear pattern.

3) Newly emerged commercial tree crop as coffee and cardamom has set spurs to various business activities, alluring many Muslim traders from neighboring villages. At the same time, a sizeable number of permanent and casual labourers has rushed in Kurubathur. Being attracted by booming plantation economy, neighbouring villagers in the Kurubathur Group *Panchayat* came to Yadavarahalli and opened up plantation estates.

4) In response to these trends in the village economy, the *patel's* house has been relocated three times up to present. His house has become closer to main settlement. Furthermore, the *Panchayat* office moved down to the present location facing main street.

These relocations as described above epitomize the changing patterns of rural settlement and ever-expanding village economy. In this sense, the cradle land of coffee adjacent to the *patel's* residence was an epoch in the development of village economy and settlement. As an important local centre, Kurubathur has thus played integral role in the Kurubathur Group *Panchayat*.

Reborn of white blossom of coffee, Kurubathur and Yadaravahalli have gradually developed into key settlements, permanently founded on the plantation economy. Village economy has been gaining momentum as it comes to the 1970s onward (Table 3–2, 3–3).

The cultivating area under the plantation crops like coffee, cardamom, ginger and others in both sample villages has been constantly growing in the recent decade. Almost half of the total cultivating land comes under the categories of plantation crop.

Thanks to full-blown plantation economy, a handful planters have enjoyed an unprecedent economic prosperity. Fifteen planters, which were selected for sample survey have almost monopolized the area under plantation crops. As for paddy, they occupy two-third of the total. Household No. K-54, K-45, K-44, K-1, K-80 and Y-3 are among the most predominant planters. The distribution of landholding is extremely polarized into the affluent big planters and the impoverished marginal class including landless, majorities of agricultural labourers.

This is partly ascribable to the fact that 'ceiling area' clauses shall not be applied to plantation estate as a rule in the Karnataka Land Reforms Act, chapter III, section 103. Most planters have enjoyed preferential ex-

Table 3–2 Crop-wise land distribution in Kurubathur Village

Category	1960-61	1964-68	1970-71	1971-72	1973-74	1977-78	1978-79	1979-80	1980-81	1981-82
Paddy cultivation land	92.39	92.39	93.59	93.59	94.95	96.37	96.37	96.37	96.37	97.00
Total tree cultivation	17.01	20.66	20.56	40.00	49.00	94.52	94.52	95.52	101.75	123.85
Cardamon wet	1.10	2.10								
Garden	2.32	2.32	3.32	3.00	3.00					
Coffee cultivable	10.24	10.24	10.24	10.00	8.00	29.19	29.19	29.19	30.19	40.19
Coffee	2.00	6.00	6.00	25.00	36.00	60.33	60.33	60.33	65.36	75.36
Orange							2.00	2.00	2.20	2.20
Plantation & garden	1.00		1.00	2.00	2.00	3.00	3.00	4.00	4.00	6.10
Coffee cultivable	0.35									
Total cultivating land	109.40	113.05	114.15	133.59	143.95	192.07	192.07	192.09	199.32	222.16
Total not cultivating land	50.64	55.52	57.62	32.18	32.82					
Out of wet (barren, fallow)	2.36	1.36	3.46	3.46	2.10					
Garden	15.01	15.01	15.01	13.33	13.33					
Coffee cultivable	33.27	39.15	39.15	15.39	17.39					
Total dry land	175.16	175.16	175.16	174.15	175.15					
Under cultivating area	9.00	9.00	9.00	22.00	23.00					
Survey free land	6.00	6.00	6.00	8.00	8.00					
Orange	3.00	3.00	3.00	10.00	10.00					
Plantation (banana)				400	500					
Not cultivating dry area	166.16	166.16	166.16	152.15	152.15					

Table 3–3 Crop-wise land distribution in Yadavarahalli

Category	1968-69	1969-70	1971-72	1973-74	1977-78	1978-79	1981-82
Paddy cultivation land	148.60	148.72	150.00	152.00	131.18	131.18	148.36
Total tree cultivation	35.98	35.96	59.79	52.68	93.55	93.55	140.06
Cardamom Wet	3.12	3.00	6.11	–			
Garden	1.00	1.00	1.00	1.00	20.35	20.35	40.36
Coffee cultivable	17.36	17.36	17.36	10.36			
Coffee	12.00	12.10	25.00	30.00	65.00	65.00	90.20
Orange	–	–	–	–	2.00	2.00	3.20
Plantation & garden*	0.30	0.30	0.32	0.32	6.20	6.20	6.30
Coffee cultivable	2.20	2.20	10.00	11.00			
Total cultivating land	184.58	184.68	209.79	204.68	224.73	288.42	
Total not cultivating land		56.69	33.08	37.23			
Out of wet (barren, fallow)		4.39	–	4.11			
Garden		0.08	–	–			
Coffee cultivable		52.22	33.08	33.12			
Total dry land		128.22	128.22	128.22	128.22		
Under cultivating area		2.21	20.0	21.00	2.21		
Survey free land							
Orange							
Plantation**		1.01	6.00	5.00	1.01		
Coffee		1.00	8.00	8.00	1.00		
Cardamom		0.20	6.00	6.00	0.20		
Not cultivating area		126.01	108.22	109.22	126.01		

* Banana, pepper, arecanuts   ** Banana, coconuts

emption clauses. The most favoured legal status of the planter protected by the Land Reforms Act have furthered vigorous reinvestment a sizeable amount of net profit in purchasing additional land suitable for plantation crops.

For further details of the recent village economy based on the plantation of coffee and cardamom, it is necessary to make a critical appraisal of well-founded statistical data, which is relevant to recent trends in the agricultural productions and its performances with reference to the agronomical aspects of plantation crops.

3. Agricultural Production

3.1 Paddy

Paddy has been cultivated as traditional main crop which provides staple food of the people. Paddy cultivation can be divided into the two types according to its farming methods, viz, dry cultivation or *'Bittane'* and wet cultivation. But in nine cases out of ten, paddy is raised in the puddled conditions. This cultivation method is locally called *'Nati'*. Productivity per acre in Kurubathur and Yadavarahalli averages 13.46 quintals with standard deviation of 30%. While the value in

Table 3–4 Productivity of paddy in Kurubathur and Yadavarahalli

House No.	Planted (acre)	Harvested (acre)	Gross production (Q)	Seed (kg)	Consumption (Q)	Amount sold (Q)	Selling price (Rs)	Productivity (Q/Acre)	Yield-seed ratio	Gross Income/acre (Rs)	Degree of commercialization	Chemical fertilizer (kg)	Organic fertilizer (kg)	Input of C.F. per acre (kg)	Input of O.F. per acre (kg)
Y-49	1.00	1.00	20	50	20	–	–	20.0	40.0	2,700.0	–	100	500	100.0	500.0
Y-4	2.00	2.00	30	100	20	10	135	15.0	30.0	2,025.0	33.3	200	5,000	100.0	2,500.0
K-81	4.20	4.20	40	100	20	20	135	8.9	40.0	1,200.0	50.0	300	2,400	66.7	533.3
Y-47	4.00	4.00	40	150	40	–	–	10.0	26.7	1,350.0	–	400	2,000	100.0	500.0
Y-45	10.00	10.00	100	400	30	70	–	10.0	25.0	1,350.0	70.0	500	50,000	50.0	5,000.0
K-51	8.00	8.00	100	300	50	50	135	12.5	33.3	1,687.5	50.0	800	10,000	100.0	250.0
Y-1	7.20	7.20	110	225	40	70	135	14.7	48.9	1,980.0		750	12,000	100.0	1,600.0
Y-6	12.00	12.00	100	600	80	20	135	8.3	16.7	1,125.0	20.0	2,000	5,000	166.7	416.7
Y-5	7.00	7.00	65	200	35	30	150	9.3	32.5	1,392.9	46.2	4,000	30,000	571.4	4,285.7
K-54	7.00	7.00	80	200	40	40	135	11.4	40.0	1,542.9	50.0	700	10,000	100.0	1,428.6
K-44	10.00	10.00	200	400	50	150	135	20.0	50.0	2,700.0	75.0	3,000	50,000	300.0	5,000.0
K-45	16.00	16.00	160	500	60	100	135	10.0	32.0	1,350.0	62.5	2,000	15,000	125.0	937.5
Y-3	18.00	3.00	50	75	–	50	135	16.7	66.7	2,250.0	74.1	1,500	30,000	83.3	166.7
		15.00	220	450	70	150	135	14.7	48.9	1,980.0					
K-1	15.00	15.00	185	400	215	185	136	12.3	46.3	1,677.3	100.0	?	?	?	?
K-80	35.00	35.00	725	800	125	600	135	20.7	90.6	2,796.4	82.8	1,500	50,000	42.9	1,428.6
U-1	28.00	28.00	400	300	60	340	135	14.3	133.3	1,928.6	85.0	2,800	42,000	100.0	1,500.0

Note : The house number do not correspond to the census numbers, but to provisional numbering for our door-to-door survey

Naravi averages 9.23 quintals with s.d. of 64.9%. This means the productivity level is not only higher, 1.46 times as much as that in Naravi, but also land productivity is more stable than in Naravi.

This is mainly due to the differences of soil fertilities and varied degrees of leaching. Degree of commercialization depends on a family size in term of consumption units. Input of chemical fertilizer per acre varies from 66.7 kg to 571.4 kg. The dose of organic fertilizer varies widely 166.7 kg to 5,000 kg.

Widely ranged distribution of organic fertilizer can be explained by the availability of cow dungs and green manures. As manures and composts can be obtained from barn, the amount of it depends on the number of livestock farmer keeps and acreage of paddy. Because paddy straws are main source of farmyard manures. Ash, tank silts, leaves are also put into use as organic manures. Nowadays available green leaves are becoming much less scarce than ever before.

The various trees as honge and athrhi are recommended to plant for the secured dose of green manures. However, the organic manures have been replaced by the chemicals since the former half of 1960s. On the other hand, ever-increasing acreage, under plantation crops as coffee and cardamom have necessitated an additional doses of organic fertilizer.

Most farmers are faced with the difficulties that they have to input diminishing amount of organic fertilizers equally to paddy and plantation crops. Some of them who found a difficulty in securing organic manures have tended to purchase it additionally from others. As they outbid each other, price of fertilizers tends to go up. Moreover, as fertile dry land and grazing land have been gradually converted into crop land, especially for plantation crops, available pasturage is becoming decreased in extent.

This process has caused over-grazing of livestock ending up to the shortage of cow dungs. Manuring problems are of primary concerns among the most planters. It might pose a serious threat to the ongoing plantation economy and the self-sufficient traditional paddy cultivation as well.

3.2 Plantation Crops
3.2.1 Coffee

India produces 100,000 tonnes of coffee out of which nearly a half amount is exported. Indian coffee earns considerable amount of foreign exchange annually. Total number of registered coffee cultivators is about 55,000, of them 95% are small growers with the extent 10ha (2.47 acres) and below. The total area under coffee is estimated to be about 160,000ha (395.360 acres), out of which Karnataka occupied nearly 60%. Coffee is cultivated as a commercial crop in the four state of South India, viz., Kerala, Karnataka, Tamil Nadu and Andra-Pradesh. It is also grown on the limited scale in Orissa, West Bental, Assam and other states.

Among these coffee growing states, Karnataka State has always been taking the lead in the production of coffee. The coffee producing areas are highly concentrated on the regions where the agro-climatic conditions are most favorable for the growth of coffee plant. As clearly seen in Table 3–5 at a glance, coffee cultivation comes to be confined to the hilly tracts of Western Ghats and Eastern Ghats. The annual rainfalls varies from 1,000 to 2,500mm.

The greater part of coffee-growing areas are under the South-west Monsoon. The plant grows well at the temperatures ranging from 15°C to 30°C.

The elevation above the sea level is decisive factor

Table 3–5 Environmental factors, economic performances and nutrient balance of coffee and cardamom

		Coffee		Cardamom
Factors \ Varieties		Arabica	Robusta	Mysore, Manjarabad, Malabar
Environmental factors	Elevation	1,000-1,500m	500-1,000m	760-1,500m
	Annual rainfalls	1,600-2,500mm	1,000-2,000mm	1,500-6,350m
	Blossom rain	March-April(2.5-4cm)	Feburary-March (2-4cm)	
	Backing rain	April-May(5-7.5cm)	April-May(5-7.5cm)	
	Shade	Medium to light shade	Thin shade	Shade regulation
	Temperature	15°C-25°C(cool)	20°C-30°C(hot and humid)	10-30°C(humid moist)
	Soils	Deep friable, porous, rich in organic matters, moisture retitive. Slightly acidic pH 6-6.5	Same as for Arabica	Well drained rich forest loams and deep light textured laterite soil abundunt in humus (not waterlogged)
	Aspect	Northern, eastern, or north-eastern aspects are most suitable	Same as for Arabica	Eastern, south-eastern
	Slope	A gentle to moderate slope	A gentle slope to fairly level field	A gentle to moderate slope
Economic performances	Year of bearing fruits	4 years after plantation	7(10)years after plantation	3-5 years after plantation
	Yields per acre	500kg	200kg	20-25kg
	Economic life of plant	40 years	100 years	15 years
	Maximum yields(15th year after plantation)	1.5 ton per acre	1.5 ton per acre	25-50 kg per acre
	Selling price (Rs) in 1972	6,000-8,000 per ton		70-100
	in 1982	10.000-12.000 per ton		100-200
Nutrient balance	Nutrient	N / P_2O_5 / K_2O	N / P_2O_5 / K_2O	N / P_2O_5 / K_2O
	Amount of nutrients absorved by one ton cream coffee(kg/ha)	34 / 5 / 45	35 / 7 / 39	
	Recommended doses of fertilizer for less than 1250 kg of crop(kg/ha)	40 / 90 / 120	80 / 60 / 80	30 / 30 / 60
	Recommended doses of fertilizer for 1250 kg of crop(kg/ha)	160 / 120 / 160	120 / 90 / 120	
	Inputs of organic fertilizer	800 kg/acre/year	500 kg/acre/year	Farmyard before planting (20 kg/plant)
	Inputs of chemical fertilizer	1.0 ton/acre/year	0.5 ton/acre/year	Manure after planting (10 kg/plant)

for the quality of coffee. Coffee *Arabica* grows well at elevation from 1,000 to 1,500m and another major type of varieties, *Robusta* grows at the lower elevations between 500 to 1,000m above the sea level. *Arabica* variety needs medium to light shade. *Robusta* needs thin shade.

As regards soil conditions both varieties grow well on a deep friable porous soils rich in organic matters and prefer northern or north-eastern aspect to south climatic and environmental factors as those mentioned above influence growth of coffee plant and quality as well. Blossam showers or backing rains from April to May and sufficient subsoil moisture are among the most important factors in the coffee plantation.

Adequate backing rains and subsoil moisture leads up to successful flowering and then to substantial bearing of fruits in due course. Excessive wetness and cool summer tends to be praticial crop failure. Careful measures should be taken against these adverse climatic and environmental limiting factors.

Brief History of Coffee Plantation

Hassan District lying on the eastern slope of the Western Ghats is one of the most favoured by these climatic and environmental factors among several coffee producing areas in Karnataka just as Nilgiris plateau in Tamil Nadu. The most important coffee growing areas extend widely over Sakleshpur *Taluk*, Belur *Taluk* and Kenchammana-Hoskote area of Alur *Taluk*. The greater part of these areas are covered with Hemavathy river system. Being well known for high-productivity of coffee as well as its scenic beauty. These areas earned national-wide repute as "Golden Land". Rolling jungles thicked with luxuriously grown bushes and the subtropical evergreen trees are very useful shade trees, which protect coffee plants from dazzling strong rays of the sun.

From the very beginning of coffee cultivation in this "Golden Land", European coffee planters had played an important role in opening up wild jungles and paved the way to the well-founded coffee kingdom just as we observe. Coffee plantation can trace back its

Table 3-6 Agricultural calendar in Kurubathur

Seasons and rain	Summer	Monsoon	Post-Monsoon	Winter
	Blossom showers*	South-west Monsoon		

Month	4	5	6	7	8	9	10	11	12	1	2	3	4

Cardamom: Transplanting (6–8), Flowering (5–8), Manuring (6–7), Manuring (9–11), Harvesting (8–12) (Mysore, Malaber, Manjarabad)

Coffee: Flowering (4–6), Harvesting (11–1) (*Arabica*), Harvesting (12–3) (*Robusta*)

Paddy: Organic fertilizer (4–5), Nursery bed (6–7), Transplanting (7–8), Chemical fertilizer (8), Weeding (9–10), Harvesting (11–12), Tilling (1–2)

* Blossom showers is also called backing rains for flowering of coffee plant.

---

history to 19 century.

A few European planters came to settle down and ventured into this remenurative business. Among those planters, Mr. Jolly of Parry and Company was the first to start coffee plantation in Karnataka. He opened up coffee estate in the neighboring district of Shimoga as early as 1823–25. Mr. Stoke was also the first man who took the leads in the eventful historical development of coffee plantation. He started this line in 1895. Lured by successful coffee plantation, several enterprizing Indian farmers joined this race one after another.

They ushered in a new era of coffee plantation. However, substantial development was made by Frederic Green on a full scale, who started coffee estate in 1843 in Manjarabad (Saklespur) *Taluk*. He was followed by Robert Elliot. When it came to 1875 number of European coffee cultivators totaled at ten. Then it had risen to twenty four, while Indian planters counted sixty. In 1946-47, the cultivated area under coffee plantation in the Hassan District was 19,062 in acreage.

In these development processes, they have learned much from down-to-earth experiment and accumulated a piece meal knowledge. On the other hand the pioneers tried to put these fragmentary know-hows together, and refined in an unified way that they could use the source of expertise as much as need be.

There had been gradual increase both in the number of coffee estates and area thereafter and a handful of pioneering European planters had been replaced by Indian planters. The number of coffee estates in Hassan District increased from 1,190 in 1948–49 to 2,170 in 1967–68.

Agricultural Calendar of Coffee Plantation

Coffee plantation needs continual care and skillful management of coffee estate for good quality and increased yields. Coffee planter's year starts with the blossom showers, which form the flowering stage to ripening usually falls in early spring. During dry season preceeding blossom showers, coffee bushes would be wintering. Soon after the first rain of blossom showers, the spikes of tiny branches begin to swell and then in ten to fifteen days, coffee trees are covered with white blossoms redolent of fragrant perfumery. (Table 3–6) At the same time planters are engaged in preparing pits of 24 inches in depth, which are arranged in rows on the terraced slopes of hills in order to make a plantation afresh. Then, in about 18 months, coffee seedlings grow enough to be transplanted in these pits from nursery beds after first rain.

Thereafter planters continue a series of works as manuring, earth up of plants, weeding, topping, prunning. Nursery trees of *Arabica* variety begin to bear fruits in the fourth year of planting. In the case of *Robusta* variety it begins to bear fruit in the seventh year of plantation.

Table 3–7 Productivity of coffee in Kurubathur and Yadavarahalli

Hause No.	Planted (acres)	Harvested (acres)	No. of plant	Yield (kg)	Consumption (kg)	Amount sold (kg)	Selling price (Rs)	Productivity (Q/acre)	Yield per plant	Gross income (Rs)	Degree of commercialization (%)	Input of C.F. (kg)	Input of O.F. (kg)	Input of C.F./acre (kg)	Input of O.F./acre (kg)	Lime* (kg)
Y–49	1.00	1.00	1,000	100	20	80	8	1.00	0.10	800	80.0	40	1,000	40.0	1,000	
Y–4	–	–	–	–	–	–	–	–	–	–	–	–	–	–	–	
K–81	4.12	–	2,775	–	–	–	–	–	–	–	–	400	30,000	93.0	697.7	1,000
Y–87	10.00	1.00	1,000	100	20	80	10	1.00	0.10	1,000	80.0	200	–	20.0	–	
Y–45	6.00	2.00	3,000	1,000	25	975	10	5.00	0.33	5,000	97.5	500	–	83.0	–	
K–51	6.000	1.00	5,500	100	25	75	10	1.00	0.02	1,000	75.0	600	15,000	100.0	2,500.0	2,500
Y–1	4.20	4.20	4,000	1,500	25	1,475	12	3.33	0.38	4,000	98.3	2,250	6,000	500.0	1,333.3	10,000
Y–6	15.00	15.00	6,000	3,000	25	2,925	10	2.00	0.50	2,000	99.2	5,000	2,000	333.0	133.0	30,000
Y–5	8.31	8.31	7,020	3,000	–	3,000	4	3.42	0.43	1,368	100.0	3,000	15,000	329.0	1,714.4	
K–4	27.00	9.00	25,000	6,000	25	5,975	10	6.67	0.24	6,667	99.6	12,000	25,000	444.0	925.9	
K–54	8.00	8.00	15,000	3,000	25	2,975	11	3.75	0.20	4,125	99.2	2,100	40,000	262.5	5,000.0	200
K–45	15.00	15.00	20,000	6,000	100	5,900	8	4.00	0.30	3,200	98.3	9,000	25,000	600.0	1,666.7	20,000
Y–3	20.00	20.00	20,000	7,000	25	6,975	12	3.50	0.35	4,200	99.6	10,000	40,000	500.0	2,000.0	20,000
K–1	5.00	5.00	850	750	250	500	5	0.15	0.88	750	66.7	n a	n a			
K–81	30.00	30.00	24,000	10,000	25	9,975	12	3.33	0.42	4,000	99.8	6,000	100,000	200.0	3,333.3	
U–1	40.00	40.00	35,000	10,000	25	9,975	12	2.50	0.29	3,000	99.8	20,000	40,000	500.0	1,000.0	

* once in every 2 years

In nearly nine months after blossoming, the berries begin to ripen. So from November to the end of February, coffee planters are very busy with harvesting crops.

Thereafter the berries are carried to pulp house where any green berries are picked out to dry separately. The ripe cherry is passed through pulpers and the skin is removed. Then the coffee beans are passed through shieves into vats. They are left there to ferment for 36 hours. After drying on the mats, they are sent to curing house at Sakleshpur or Hassan.

Economic Performance

Arabica variety could be divided into six commercially important selections; 1) Old Chicks, 2)Coorgs, 3) Kents, 4) S.288, 5) S.795, 6) S.1934. Old Chicks were originally grown in Chikmagalur and Coorg seem to have originated in Naraknad area of Coorg. It came under cultivation in 1870. Kent reliction was developed by L.P. Kent of Doddangudda and Kalagnee estates in Mudigere. This variety came under cultivation in about 1918–20 and shortly spreaded widely over India because of its high yielding, with standing higher temperature and tolerance to leaf-miners. Old Chicks variety is famous for its high liquoring quality. Coorgs is characterized by graceful drooping margins, but susceptible to leaf rust just as Old Chicks.

Arabica varieties come to bear fruits 4 years after plantation and yields 500kg per acre. In nearly fifteen years after plantation, yield reached at its peak to the level of 1.5 ton per acre. Economic life of plant spans over 40 years. As regards *Robusta* variety, it is bigger than *Arabica* with broader and larger leaf, coloured pale green. Flower are white and fragrant. This variety come to bear fruits in seven years after planting. It yields 200kg per acre on the tenth year and yields goes up to 1.5 ton per acre in the fifteenth year after planting. Economic life of the *Robusta* variety spans over 80 to 100 years.

Selling price of each varieties was 6,000 to 8,000 Rs per ton in 1972. It has come to the level ranging 10,000 to 12,000 Rs (Table 3–5). As perennial coffee trees, once planted, secure a constant series of annual incomes, land price rose up to 5,000 Rs per acre in 1982 as against 1,000 Rs in 1975 (Table 3–5).

Seeing from the nutrient balance, revealed in the amount of nutrients absorbed by one ton clean coffee (kg/ha) and recommended doses of fertilizer, in the ease of coffee *Arabica* yielding less than 1,250 kg of crop, the amount of N is nearly balanced. But inputs of P_2O_5 are greatly needed more than that of K_2.

When it comes to coffee plants yeilding 1,250 kg and more, planters have to input 160 kg of N and K_2O, 4 times greater than the amount of nutrients removed by one ton of clean coffee. The same thing could be said as for *Robusta* variety except P_2O_5. On the average *Robusta* variety is economical from the veiw point of inputs. Inputs of organic fertilizer for *Arabica* average at 800 kg per acre as against 500 kg for *Robusta* and chemical fertilizer 1.0 ton per acre for *Arabica* as against 0.5 ton per acre for*Arabica*.

The lost nutrients by the crop account for only one-third to fourth of the total requirements needed for successful yields. The gap of nutrient balance could be explained by leaching under the tropical heavy rainfalls. This leads to an improverishment of soils and consequent decrease in yields.

Therefore it is necessary to make up for the last nutrients by applying additional fertilizers sufficient to a secured annual yields. In fact, as clearly shown in Table 3–7, as for the amount of inputs of chemical fertilizer per acre, most of planters apply 400–500 kg except a few cases. The inputs of organic fertilizer varies from 133 kg to 5,000 kg. Most of them input more than 1 ton per acre along with lime.

Table 3–8 Productivity of cardamom and chilli in Kurubathur and Yadavarahalli

House No.	Planted (acre)	Harvested (acre)	No. of plant	Yield (kg)	Consumption (kg)	Amount sold (kg)	Selling price (Rs)	Productivity (Q/acre)	Yield per plant	Gross income	Degree of commercialization (%)	Input of C.F. (kg)	Input of O.F. (kg)	Input of C.F. /acre (kg)	Input of O.F. /acre (kg)
Cardamom															
Y–49	0.20	0.20	500	20	–	20	100	40	0.040	4,000	100	–	–	–	–
Y–4	2.00	1.00	3,000	20		20	100	20	0.006	2,000	100	–	–	–	–
K–81	–	–	–	–	–	–	–	–	–	–	–				
Y–47	1.00	1.00	1,000	10		10	100	10	0.010	1,000	100	20	–	20.0	–
Y–45	3.00	2.00	2,000	10	–	10	100	5	0.005	500	100	200	–	66.7	–
K–51	4.00	4.00	4,000	200		200	100	50	0.050	5,000	100	150	–	37.5	–
Y–1	3.00	3.00	3,000	75	3	72	100	25	0.025	2,500	96	200	5,000	66.7	1,666.7
Y–6	3.00	3.00	3,000	150	5	145	100	50	0.050	5,000	97	1,000	3,000	333.3	1,000.0
Y–5	6.00	6.00	9,000	150	–	150	100	25	0.017	2,500	100	500	–	83.3	–
K–54	1.13	1.13	1,500	50	2	48	100	37.7	0.033	3,775	96	250	2,000	188.7	1,509.4
K–44	12.00	12.00	10,000	150	5	145	100	12.5	0.015	1,250	97	200	20,000	16.7	1,666.7
K–45	4.00	4.00	10,000	100	3	97	100	25	0.010	2,500	97	300	50,000	75.0	1,250.0
Y–3	3.00	3.00	3,500	150	5	45	100	50	0.043	5,000	30	500	6,000	166.7	2,000.0
K–1	8.00	8.00	1,250	200	–	200	80	25	0.160	2,000	100	n a	n a	n a	n a
K–80	10.00	10.00	15,000	350	5	345	100	35	0.023	3,500	99	500	800	50.0	80.0
U–1	12.00	4.00	10,000	100	5	95	100	25	0.010	2,500	95	–	1,200	–	10.0
Chilli															
Y–49	0.10	0.10	2,000	100	50	50	12	400	0.05	4,800	50				
Y–45	1.20	1.20	1,000	100	30	70	10	66.7	0.10	667	70				

According to the village survey monographs (Salasubramanyam, 1961), the area under coffee in Kurubathur village was enumerated to be only 5 acres and cultivated by the two planters in the 1950s. Coffee was a minor product, most of which was used for home consumption.

But in the recent decade, the acreage of coffee plantation has jumped up to 80.10 acres in this village and 106.01 acres in Yadavarahalli. A phenomenal increase of the acreage under coffee is mainly due to its high profitabilities. But judging from agro-climatic point of view, coffee plant can better withstand untimely heavy rainfalls and can be resistent to strong sunlights. It need not so much skillful and constant cares as cardamom. These factors have furthered coffee plantation and taken the place of cardamom.

3.2.2 Cardamom

Cardamom (*Elettaria cardamomum*) is regarded as the 'Queen of Spices'. There are 35 spices and condiments, which could be broadly classified into 6 groups; 1) rhizomes and root spices, 2) bark spices, 3) leaf spices, 4) flower spices, 5) fruit spices, and 6) seed spices. Cardamom falls into the category of fruit spices and a perenial herbaceous plant belonging to the family *Zingiberaceae*.

The plant begins to bear fruits in 3 to 5 years after planting. A normal full-grown plant stands at 2 to 4 meters in height. Economic life of plant usually spans over 15 years. Annual yields per acre are estimated to be 25 to 50kg. Varied are the uses of cardamom. However, dried fruits of cardamom are mostly used as additives for stimulating digestion on account of their carminative properties. The plant of cardamom has been growing in India from time immemorial and native of India. It is one of the most important commercial crops grown on a large scale, along with other spices like pepper, chillis, turmeric and ginger. In this area, ginger is planted as a side crop of coffee plantation.

India is the largest producer and exporter of cardamom, accounting for nearly 70 per cent of the total world production and 60 per cent of the total world trade. Annual production of cardamom is estimated to be nearly 3,000 tonnes in recent years. This plants are cultivated in the hilly regions of the entire Western Ghats in Karnataka, Kerala, Tamil Nadu on about 90,000 hectares. The varieties are classified into the four types, viz, Mysore, Manjarabad, Malabar and Vazukka.

Environmental Factors of Cardamom Plantation

Cardamom grows well in the tropical evergreen forests of the Western Ghats. It thrives best in tropical forest at altitudes ranging from 750 to 1,000m, receiving an evenly distributed rainfalls of over 1,500mm per annum. The most favorable temperature for the growing of the plant ranges widely 10°C to 30°C. It is highly responsive to wind and excessive moisture are equally unsuitable to the plant.

Cardamom is grown on well-drained rich forest loams and deep light textured laterite soils, rich in humus (Table 3–5). Preferable aspects is northern and eastern or north-eastern. In the case that cardamom is grown as a subsidiary crop of arecanuts or coffee, the main crop furnishes the required shade.

The spacing of plants commonly adopted varies from 1.5m to 3.3m between the plants and the rows. The readlings are planted in pits of about 60cm × 60cm × 35cm filled with a mixture of surface soils rich in humus and midden composts. cardamom is propagated

Table 3–9 Distribution of farming size and landholding of the sample household

House No.	*Jati*	Paddy	Coffee	Cardamom	Grazing land	Waste land	Total acreage
Y–49	*Vokkaliga*	1.00	1.00	0.20	1.00		3.20
Y–4	*Lingayat*	2.00		2.00	6.00		10.00
K–81	*Lingayat*	4.20	4.12		7.00		15.32
Y–47	*Vokkaliga*	4.00	10.00	1.00	1.00		16.00
Y–45	*Vokkaliga*	10.00	6.00	3.00		6.00	25.00
K–51	*Vokkaliga*	8.00	6.00	8.00		4.00	26.00
Y–1	*Lingayat*	7.20	4.20	3.00		16.20	31.20
Y–6	*Lingayat*	12.00	15.00	3.00	3.00		33.00
Y–5	*Lingayat*	7.00	8.31	6.00	14.00		35.31
K–54	*Vokkaliga*	7.00	27.16	1.13	3.00		38.29
K–44	*Lingayat*	10.00	8.00	12.00	12.00		42.00
K–45	*Lingayat*	16.00	15.00	4.00	7.20		42.20
Y–3	*Lingayat*	18.00	20.00	3.00	32.00		73.00
K–1	*Lingayat*	15.00	25.00	8.00	30.00		78.00
K–80	*Lingayat*	35.00	30.00	10.00	40.00		115.00
U–1	*Lingayat*	28.00	40.00	12.00	90.00	20.00	190.00
Total		184.4	219.79	76.33	246.2	46.20	773.52
Total excluded House No. U–1		156.4	179.79	63.44	156.2	26.20	583.52

Note: Estimated land price per acre and price of a pair of bullock in Kurubathur (Rs).

	Paddy	Coffee	Cardamom	Grazing land	A pair of bullock
1972	4,000	1,000	300	50	3,000–4,000
1982	10,000	5,000	2,000	2,000	5,000

through suckers on seedlings. However, the seedling method is widely used among the cultivators because it checks the spread of urius disease "Katte". The selection of the site for the nursery bed is very important to ensure uniform and vigorous growing young plants.

The site for nursery bed should be well irrigated with gentle slope. But, in the sample surveyed villages, nursery beds are usually prepared amidst paddy fields. The shabby shacks are built up on the heightened nursery bed to prevent seedlings from strong wind and sunlight. Germination will begin in a month. The mulch should be put off after germination.

Shade has to be provided for the young seedlings by erecting pandal seedlings are grown in the bed for 10 months. Thereafter they are transplanted in the main field. One to two-year-old seedlings are planted in each pits either in June-July or in September-October.

Cardamom comes into bearing in the third to fifth year after planting. However, sizeable amount of crop can be obtained from that year onwards. Flowering begins in April to May and continues to August to September. Harvesting season starts in July-August and continues up to December-January (Table 3–6). Yields per acre ranges from 25 to 50kg depending on soil fertilities amount of input of fertilizer, controlling of shade and water (Table 3–8). Selling price was 70 to 100 Rs in 1972 and has risen up to about 200 Rs in 1982. Fertilizer is recommended to apply for the healthy and vigorous growth of plant at the rate of 30kg(N); 30kg (P_2O_5); 160kg (K_2O) per acre equally first in May-June and the second in September-October.

The cultivation of cardamom involves lots of hard works as skillful and constant cares for mulching, water control, shade control and so on. Furthermore annual yields tends to fluctuate widely and therefore selling price is also covariant with productivity level. The plant of cardamom is very subject to strong wind and sunshines.

Capricious onset of monsoon is among the most decisive hazards. In this sense, the cultivation of cardamom is profitable but very risky adventure. The planter cannot make money unless they are well prepared to run the triple risks of losing money, times involved in careful treatment of the plant and of exhausting a limited amount of soil-fertility.

The plant was first cultivated by European planters on the virgin soil rich in forest humus at the beginning of this century. They could get a sizable amount of yields. However, successive cropping of the plant year after year had caused a gradual decrease in yields because of deterioration and exhaustion of soil fertilities. Cardamom planters had been forced to single out the alternative to give up growing Cardamom or to introduce shifting cultivation. As of 1961, it was restricted to 75 acres. So the cardamom came to minor crops cultivated by the ten planters. Thus cardamom has been replaced largely by coffee plantation in these recent two decades.

3.2.3 Gross Incomes

In order to clearly demonstrate economic perform-

Table 3–10 Farming size and economic performances of farm management in Kurubathur and Yadavarahalli

House No.	Y–49	Y–4	K–8	Y–47	Y–45	K–51	Y–1	Y–6	Y–5	K–54	K–44	K–45	Y–3	K–1	K–80	U–1
Landholding	3.20	10.00	15.32	16.00	25.00	26.00	31.20	33.00	35.31	38.29	42.00	42.20	73.00	78.00	115.00	190.00
Agriculture	2,000	4,000		5,000	25,000	40,000	50,000	60,000	60,000	100,000	80,000	125,000	125,000	28,000	125,000	180,000
Cattle breeding																
Land rent																
Casual agricultural labourer																
Business			3,600	500	2,000											
Working away from home																
Other																
Total	2,000	4,000		5,000	25,000	40,000	50,000	60,000	60,000	100,000	80,000	125,000	125,000	28,000	125,000	180,000
Chemical fertilizer	400	270	4,000	1,500	5,000	3,900	8,000	10,000	8,000	32,000	23,000	20,000	27,000	6,500	52,000	60,000
Organic fertilizer	200	400	2,000	1,000	2,000	1,500	2,000	4,000	1,500	4,000	5,000	500	6,000	400	5,000	6,000
Insecticide, herbicide	200	200	500	500	200	250	250	500	2,000	3,500	560	500	4,000	250	8,000	10,000
Seeds and young plants	100	300	2,000	200	500	500	500	500	500	300	1,000	2,000	1,500	1,000	5,000	2,000
Animal feeds	100		2,500	200	500	1,000	1,000	1,000	2,000		1,500	1,500	3,000	1,500	1,500	5,000
Fuel and electricity										2,500						
Payments to labourers	300		3,500		4,000	4,000	6,000	25,000	7,000	50,000	12,000	20,000	30,000	15,000	18,000	70,000
Land rent																
Costs for Irrigation					200			800	700	1,200	200	300	200	700	500	1,000
Transportation costs	50	50		50	500	700	500	400		2,000	150	500	3,000	4,000	1,000	2,000
Purchase of small animal																
Purchase of large animal					4,000	1,500				1,600	3,000		1,500		2,500	3,000
Purchase and repairment of faming tools and machines						250	5,000						13,000			35,000
Land tax	100	200	75	275	600	350	200	400	500	400	625	1,500	570	550	2,000	
Other farm expenditures			1,000							3,500				2,500		
Total farm expenditures	1,450	1,420	15,575	3,725	17,500	13,700	18,700	47,600	27,300	98,900	44,035	46,800	89,770	32,400	95,500	194,00
Monthly living expenditures of household	200	500	1,000 1,500	500	500	1,200	1,000	1,000	1,000	2,000	800	1,500	2,000	1,500	1,000	2,500
Loan																
Amount	–	–	15,000	2,000	3,000	2,000	12,000 15,000	25,000	25,000	80,000	20,000	–	60,000	25,000	30,000	140,000
Interest rate pa(%)			12.1	13.5	8.1	18.0	12.5	12.0 18.0	13.5	12.5	12.5			10.0	10.0	14.5
Purposes* : 1st			Cp	Cp	Cp		Fe	Li	Pa	Pa	Pa		Pg	Pa	Pa	Li
2nd			Pa	Fe	Pa			Fe	Fe	Fe	Fe		Fe	Fe	Fe	Fe
3rd			Fe													
Gross incomes per acre	571.4	400.0		312.5	1,000.0	1,538.5	1,587.3	1,818.2	1,677.1	2,582.3	1,904.8	2,941.2	1,712.3	359.0	1,087.0	947.4
Gross incomes per acre **	800.0	1,000.0		333.3	1,315.8	1,818.2	3,333.3	2,000.0	2,755.5	2,799.2	2,666.7	3,571.4	3,048.8	583.3	1,666.7	2,250.0

* Cp; Coffee plant, Pa; Payment, Fe; Fertilizer, Li; Land improvement
** Excluded dry land and grazing land

ances of farm managements in the sample villages, the detailed input-output balances are summarized in Table 3–10. The sixteen sample households are arranged according to their landholding size (Table 3–9). Roughly estimated annual gross incomes are covariant with landholding size. The gross incomes derive mainly from plantation by income size, ie, marginal class with annual incomes less than 5,000 Rs, lower middle class with annual incomes ranging from 25,000 to 50,000 Rs, upper middle class from 50,000 up to 100,000 Rs, and large sized class above 100,000 Rs (Table 3–9, 3–10).

The absolute incomes increase according as landholding size goes up. However, gross incomes per acre tend to increase constantly up to the middle classes, reaching its peak at the household Y-6, thereafter, alternating between ups and downs and then decrease to the level of 1,000 Rs.

But supposing that growing land and waste land does not contribute directly to annual gross incomes it goes up to the level ranging from 2,000 to 3,000 Rs except few cases.

Turning to farm expenditures, chemical fertilizer, organic fertilizer, insecticide, pesticide, payments to labourer and purchase and repairement of tools & machines are among the major items of expenditures. Especially, chemical fertilizer is one of the most deteriorating factors for the total balance of agricultural management in the sample village.

Reference

Balasubramanyam, K. ed. (1970): Census of India 1961, XI (Mysore), Village survey Monographs, No. 15 (Kurubathur village). Delhi, Govt. of India, 73p.

CHAPTER 4
AGRICULTURAL LAND UTILIZATION AND PRODUCTIVITIES IN NARAVI

1. Agricultural Land Utilization

The geographical area of Naravi Village is 6,781.45 acres and the total extent of land is divided into two categories, viz, agricultural land and forest land in which the afforested area is included. The agricultural land is 3,142.43 acres (46.3%) as much as nearly forest land with the extent of 3,239.02 acres (47.8%). Most of the eastern part of the village forming steeper slope of Western Ghasts is now thickly covered with the sub-tropical evergreen forests owing to a heavy rainfalls. (Fig. 4–1).

The sample survey had been made mainly on the eastern part of the village which is composed of hill region, river terraces and alluvial plains. The total extent of agricultural land is classified into four categories, viz, wet land (11.6%), garden (1.27%), dry land (27.58%) and cultivable land (5.89%).

Among, the four categories of land use, it is interesting to note that the extent of dry land amount nearly one third of agricultural land. As of the significance of dry land in the agricultural activities shall be minutely examined in the next section. Wet land is vital importance in the village economy. The whole area of wet land is classified into three classes; wet I, wet II, and wet III. Graded evaluation of wet land is mainly based on the availabilities of water and economic soil capabilities assessed by the land revenue office. Wet I is most favourable in the availability of water, making best use of amples rainfalls of the South-west Monsoon, followed by post-monsoon. On this paddy field of Wet I, the frequencies of land use extend over double cropping or even to triple cropping. Class-wise cropping patterns and their average productivities are indicated in Table 4–1. One cropping of rice on Wet I through III classes is locally called as *yenel*, double cropping on Wet I as *suggi*, and triple cropping as *kolake*. Agricultural calender of paddy is illustrated in Fig. 4–2.

From those facts it comes to clear that the cultivation of paddy are primarily concerned with the availabilities of water, which is heavily dependent on t rainfalls by South-west Monsoon. In this sense, farming system of rice cultivation in this village could be classified as rain-fed paddy. However in the double and triple-cropped paddy, some patches of paddy fields are mainly irrigated by tank system. But the water supply of a irrigation is limited in its capability. It is often pointed out that one tank can afford to supply water over the extent of 5 acres of paddy.

2. Productivities and Circulation of Agricultural Reproduction

In order to clearly demonstrate the intensity and economic performances of the paddy field, Table 4–4 was tabulated, showing total balance of inputs and outputs at a glance. An intensified land use can be seen in

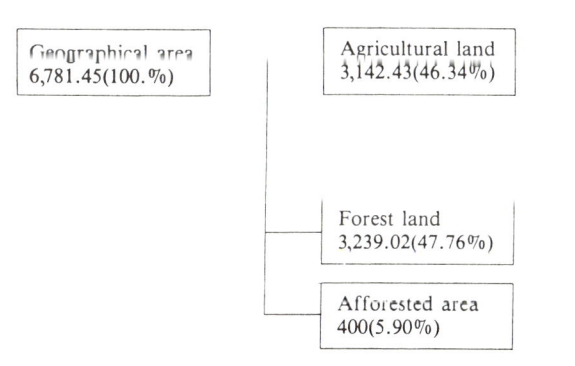

Geographical area 6,781.45(100.%)	Agricultural land 3,142.43(46.34%)	Wet land 786.19(11.6%)	Wet I 324.38(4.79%)
		Garden 86.37(1.27%)	Wet II 127.41(1.88%)
	Forest land 3,239.02(47.76%)	Dry land 1,869.87(27.58%)	Wet III 305.60(4.51%)
	Afforested area 400(5.90%)	Cultivable waste 400(5.89%)	Others 28.6(0.42%)

Fig. 4–1 Categories of land use and acreages in Naravi

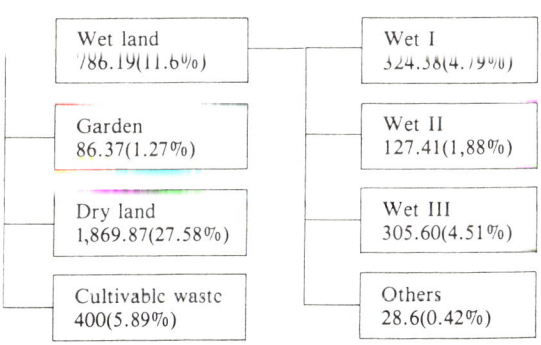

Table 4–1 Cropping patterns and yields of class-wise wet lands in Naravi

(Quintal/acre)

Wet land	Cropping patterns and yields		
	1st cropping	2nd cropping	3rd cropping
I	9	9 *(suggi)*	10 *(kolake)*
II	9 *(yenel)*		
III	8-7		

Month	Mar.	Apr.	May	June	July	Aug.	Sept.	Oct.	Nov.	Dec.	Jan.	Feb.	Total
Rainy day	0	3	6	28	28	31	21	10	5	1	0	0	133
Rainfall in mm	—	52.0	98.0	1159.0	1231.0	1461.0	558.0	124.0	92.0	4.0	—	—	4779
Rainfall in %		1.09	2.05	24.25	25.76	30.57	11.68	2.59	1.92	0.08	—	—	100.0
(Cumulative%)		(1.09)	(3.14)	(27.39)	(53.15)	(83.72)	(95.40)	(97.99)	(99.91)	(100.0)			

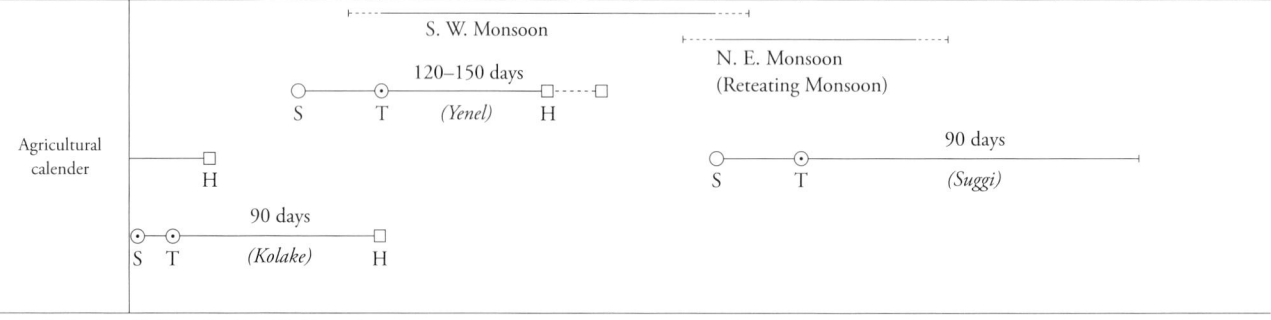

Agricultural calender:
S. W. Monsoon, 120–150 days, S T (Yenel) H
N. E. Monsoon (Reteating Monsoon), 90 days, S T (Suggi)
90 days, S T (Kolake) H

S : Sowing, T : Transplanting, H : Harvesting

Fig. 4–2 Agro-climatic conditions and agricultural calender of rice cutivation in Naravi

Table 4–2 Distribution of sample landholdings classified by land use categories in Naravi, 1982

(Acre-Gunte)

House No.	Jati	Agricultural land							
		Irrigated				Unirrigated			
		Wet I	Wet II	Wet III	Total	Garden	Dry	Waste	Total
1–1	*Billawa*	2.00*			2.00		2.00		4.00
4–2	Jain		2.20		2.20	0.20	1.00		4.00
4–3	*Billawa*	2.00	2.00		4.00				4.00
4–172	G.S.B.						4.20*		4.20
3–4	*Billawa*	2.00	2.00		4.00		1.00		5.00
2–149	G.S.B.		3.00				1.00*	1.00	5.00
1–51	Brahmin		1.20		1.20		4.00		5.20
2–86	Jain	3.00	0.20		3.20	0.20	2.00		6.00
1–74	Christian	1.00	1.00	0.20	2.20		4.00		6.20
2–166	*Billawa*	6.00			6.00		0.30		6.30
2–140	Jain	3.00	1.20		4.20	0.20	2.20		7.20
2–5	Jain	4.00	3.00		7.00		2.00		9.00
2–6	Jain	5.00	1.20		6.20	0.20	2.00		9.00
1–71	*Billawa*	3.00	3.00	1.00	7.00		3.00		10.00
2–138	Jain	3.20	1.00	2.00	6.20	0.20	2.00	1.00	10.00
2–7	Jain	4.20	2.00		6.20		4.00		10.20
2–38	Christian	5.25	2.00	2.00	9.25	0.10	1.00		10.35
3–190	Brahmin	0.20			0.20	8.00	3.00		11.20
1–18	*Heggade*	5.00	3.00	0.15	8.15	0.20	4.00		12.35
2–8	*Billawa*	4.00	4.00		8.00	1.30	4.00		13.30
4–9			3.00	3.00	6.00	3.00	6.00		15.00
1–10	*Billawa*	4.00	3.00	2.00	9.00	2.00	8.00		19.00
2–142	Christian	4.00	5.00	1.20	10.20	1.00	8.00		19.20
2–87	Jain		3.20	9.00	12.20	2.00	6.00		20.20
1–16	*Billawa*	7.00	6.00		13.00	0.20	12.10		25.30
Grand Total		70.85	51.12	17.55	138.26	17.180	86.80	2.00	248.32
(%)		72.05	54.00	18.15	144.20	21.20	88.00	2.00	256.00
		28.17	21.10	7.17	56.44	8.40	34.38	0.78	100.00

Note : The house numbers do not correspond to the census numbers, but to provisional numbering for our door-to-door survey.

the present multiple cropping pattern of paddy. From the economic point of view, the intensity of agricultural land use could be measured not only by the frequency of land use but also the achieved performances in term of outputs per a unit of land.

The author selected out 25 farmers as a sample household which were willing to give us reliable informations on the detailed managerial data, considering their representativeness of the stratified landholding sizes, varieties of major crops, and farming orientation into account (Table 4–3). The results are summarized in Table 4–4 and Table 4–5. Analysis have revealed interesting facts on the economic performances which is expressed as quintals per acre. Table 4–5 shows the frequencies distribution of productivities of paddy in Naravi as of Sept.1982.

For the convenience of cross-reference to the another sample survey village, the results of Kurubathur and Yadavarahalli have been combined. As for Wet I, it gained 9.11 Q/acre. As standard deviation is 5.80, the variation to the average value stands relatively high. When it comes to Wet II, outputs is rather high. But basically productivity level remains unchanged. Wet III was caluculated to be 8.40. Judging from the fact that the coefficient value almost the same with the 60 percent level, variations to each average ($\bar{x}$) is to be equally interpretated. Productivities of paddy stand at the level of 9.2 Q/acre. The results of Kurubathur and Yadavarahalli is estimated to be 13.5 Q/acre, 1.46 times greater than that of Naravi. 15 households with the value less than 5 Q/acre amount the nearly one third out of all the samples. These are far below the level and could be termed as an extremely low productivity land. There lacks the corresponding class in the case of Kurubathur and Yadavarahalli. It is also bassically true of an other indicator showing economic performances of land use, yields-seeds ratio, that performance level is considered to be substantially lower than that of counterparts. Yields-seeds ratio in Naravi come to be around 21 Q/acre on the whole. In Kurubathur and Yadavarahalli it goes up to 47 Q/acre, it is found to be 2.24 times greater than Naravi.

In addition in Naravi frequency distribution of productivities varies widely from 3 Q up to 75 Q/acre. This resulted in raising up Q. As for the case of Kurubathur and Yadavarahalli, dividing line is estimated to be 40 Q and 90 Q per acres and sample population is also relatively uniform than Naravi.

Careful attention should be focused on the various aspects of productivity gap between two sample villages. Among many a plausible reasons, low productivities revealed in both quintals per acre and yield-seeds ratio, seem to be closely concerned with the ecological dimension of the subtropical region. In recent years, there have been increasing number of farmers, introducing hybrid varieties as seen in Table 4–4 instead of local seeds. Hybrid varieties require much of chemical fertilizer as well as organic fertilizer in appropriate proportion to it.

The facts that widely ranged distribution of productivities and its polarized concentration either to lower and higher class can be partly explained an availabilities of chemical fertilizer. Uneven inputs of chemical fertilizer as indicated in Table 4–4 confirm this discussion. If they amply input chemical fertilizer along with midden compost, and increase as much as paddy field palnts permit to the reasonable limits, much of it shall be washed away to down streams by the continued heavy rainfalls in the monsoon season. Diurnal showers in every afternoon tend to accelarete "carry-away" of nutriments rich in fertilizers. Farmer complains that, fertilizer once applied to "this own" land, it shall go away with water to somewhere, and there is no way to get it restored. Amount of run-offs of fertilizer caused by the diurnal torrential showers and continued monsoon as well should not be disregarded in this respects. Heavy rainfalls in the subtropical region like a village of Naravi, which have been thought to be rather advantageous and favourable not only to rice cultivation, but for the purpose of versatile use of fully grown natural vegitations. It is admittedly appropriate to say that heavy rains is a matter of balancing profits against losses, what we call swings and roundabouts.

Another major factor responsible for the low level of productivities are considered to be a soil type and its fertilities. Generally speaking, most of the land in the village is composed of red loam and laterite, interspersed with sandy loam soils. Paddy of sandy loam soils do not retain nutriments so much as expected. The degree of soakage is comparatively high. According to the detailed examination of soil structures in this village indicates that soil itself is of acid and infertile. This is all the more reason to cultivate those types of land as a paddy. Paddy plants are resistant to acid soils and less susceptibe to soil erosion.

However profitability of paddy cultivation has turned out to be worse in recent years because chemical fertilizer, labour cost and maintenance costs of machinary as tiller or pompset have soared up. Pay-off balance of agricultural management is levelling off. Paddy has ceased to be commercial crop and is now grown for mainly home consumption.

Degree of commercialization ranging from the level of 30% at most to 9.3% validates this view (Table 4–4). Out of 25 households, 10 households cultivated paddy only for home consumption. Most of them have tended to diversify the variety of major crops than ever before. This tendency are apparently traced on the var-

Table 4–3 Productivities of paddy in Naravi

House number	Wet I Acreage Planted	Wet I Acreage Harvested	Gross production	Seed (kg)	Productivity (Q/Acre)	Yield-seed ratio	Gross-Income per acre
1–1							
4–2							
4–3	2.00	2.00	20	80	10.0	25.0	1,292.0
4–172							
3–4	2.00	2.00	12	80	6.0	15.0	840.0
2–149							
1–51							
2–86	3.00	3.00	30	400	10.0	7.5	1,300.0
1–74	1.00	1.00	9	80	9.0	11.3	1,162.8
2–166	6.00	6.00	40	300	6.7	13.3	866.7
2–140	3.00	3.00	(36)	(400)	8.0	9.0	1,033.6
2–5	4.00	4.00	30	200	7.5	15.0	969.0
2–6	5.00	5.00	12	400	2.4	3.0	310.1
1–71	3.00	3.00	15	120	5.0	12.5	646.0
2–138	3.00	3.00	(75)	(100)	18.8	75.0	2,343.8
2–7	4.20	4.20	25	150	5.6	16.7	694.4
2–38	5.25	5.25	40	160	7.1	25.0	888.8
3–190	0.20						
1–18	5.00	5.00	20	240	4.0	8.3	520
2–8	4.00	4.00	20	100	5.0	20.0	625
4–9	3.00	3.00	(140)	(300)	23.3	46.7	3,014.7
1–10	4.00	4.00	30	160	7.5	18.8	975
2–142	4.00	4.00	(200)	(800)	19.0	25.0	2,380.9
2–87	—	—	—	—			
1–16	7.00	7.00	40	250		16.0	738.3

House number	Dry land rice Amount seed	Dry land rice Selling price (Rs)	Dry land rice Degree of commercial-ization	Wet I Chemical fertilizer	Wet I Organic fertilizer	Wet I Input of C. F. per acre	Wet I Input of O. F. per acre
1–1							
4–2	0	—	0				
4–3	0	—	0	200	20,000	100	10,000
4–172	0	—	0				
3–4	0	140	0	—	20,000		10,000
2–149	7	140	9.3				
1–51	0	—	0				
2–86	13	130	37.1	—	40,000		13,333
1–74	0	—	0	20	10,000	20	10,000
2–166	10	130	25.0	—	30,000		5,000
2–140	0	—	0	300	10,000	10	3,333
2–5	0	—	0	—	20,000		5,000
2–6	0	—	0	—	30,000		6,000
1–71	5	130	14.3	—	30,000		10,000
2–138	20	125	26.7	150	2,000	50	667
2–7	10	125	28.7	450	25,000	100	5,555
2–38	30	125	37.5	200	25,000		4,444
3–190		—					
1–18	5	130	16.1	400	50,000	80	10,000
2–8	5	125	14.3	—	25,000		6,250
4–9	0	—	0	500	5,000	83	833
1–10	20	130	33.3	200	30,000	50	7,500
2–142	40	125	20.0	200	25,000	50	6,250
2–87	35	125	33.3				
1–16	0	—	0	—	30,000		4,286

Chap.4 AGRICULTURAL LAND UTILIZATION AND PRODUCTIVITIES IN NARAVI

Wet II						
Acreage		Gross production(Q)	Seed (kg)	Productivity (Q/Acre)	Yield-seed ratio	Gross-Income per acre
Planted	Harvested					
2.20	2.20	30	100	12.0	30.0	1,550.4
2.00	2.00	20	80	10.0	25.0	1,292.0
2.00	2.00	12	80	6.0	15.0	840.0
3.00	3.00	75	150	25.0	50.0	3,500.0
1.20	1.20	5		3.3		430.7
0.20	0.20	5	30	10.0	16.7	1,300.0
1.00	1.00	9	80	9.0	11.3	1,162.8
1.20	1.20	(36)	(400)	8.0	9.0	1,033.6
3.00	3.00	20	150	6.7	13.3	861.3
1.20	1.20	4	50	2.7	8.0	344.5
3.00	3.00	15	120	5.0	12.5	646.0
1.00	1.00	(75)	(100)	18.8	75.0	2,347.8
2.00	2.00	10	75	5.0	13.3	625.0
2.00	2.00	20	50	10.0	40.0	1,250.0
3.00	3.00	10	150	3.3	6.7	433.3
4.00	4.00	15	100	3.8	15.0	468.8
3.00	3.00	(140)	(300)	23.3	46.7	3,014.7
3.00	3.00	20	120	6.7	16.7	866.7
5.00	5.00	(200)	(800)	19.0	25.0	2,380.9
2.20	2.20	(105)	(800)	9.1	13.1	1,141.3
6.00	6.00	30	200	5.0	15.0	646.0

Wet II				Wet III			
Chemical fertilizer	Organic fertilizer	Input of C. F. per acre	Input of O. F. per acre	Chemical fertilizer	Organic fertilizer	Input of C. F. per acre	Input of O. F. per acre
100	2,000	40	800				
–	20,000		10,000				
–	25,000		12,500				
300	3,000	100	1,000				
–	1,000		667				
–	20,000		40,000				
20	10,000	20	10,000	2,000	10	4,000	20
100	5,000	67	3,333	–	–		
–	15,000		5,000				
200	5,000	133	3,333				
–	30,000		10,000	–	5,000	–	5,000
50	2,000	50	2,000				
200	10,000	100	5,000				
100	10,000	50	5,000	100	10,000	50	5,000
200	30,000	67	3,333	50	1,000	133	2,667
100	20,000	25	5,000				
25	500						
–	25,000		8,333	100	10,000	50	5,000
100	30,000			50	8,000	33	5,333
300	15,000			900	40,000	100	4,444
100	20,000						

Table 4–3 (Continued)

	Wet III					
Acreage		Gross production	Seed (kg)	Productivity (Q/Acre)	Yield-seed ratio	Gross-Income per acre
Planted	Harvested					
0.20	0.20	4	40	8.0	10.0	1033.6
1.00	1.00	5	30	5.0	16.7	650.0
2.00	2.00	20	50	10.0	40.0	1250.0
0.15	0.15	1	10	2.7	10.0	46.7
2.00	2.00	10	80	5.0	12.5	650.0
1.20	1.20	(200)	(800)	19.0	25.0	2380.9
9.00	9.00	(105)	(800)	9.1	13.1	1141.3

Seeds						
Local Varieties			High Yielding Varieties			Dry land rice
Wet I	Wet II	Wet III	Wet I	Wet II	Wet III	
Kavalagarru	(Rajakayame / Kayame)		Masuri	Jaya		
Kayame	Suggi Kayame			Jaya		Maradi
Ganjasale						
(Kamalakanni / Kayame)	(Masuri-Ni ranbde)			Jaya		
(Rajakayame / Kumavabede)		(Kumavabede / Shakthi)		Masuvi		
Kayame	Rajakayame			(Jaya / I R-8)		
Kayame	Rajakayame					
(Doddara / Kayame)	Athikara					
(P. T. B / Kavalagerru)	Pottamba Kayame		Shakthi			
(Kayame / Rajakayame)	J. B	(Rumave / Bidar,)	Jonur			
(Kavalakarru / Kayame)	Doddara J. B		(Shakthi / Jaya)			
(Kavalakarru / Kayame)	Athikara			(Shakti / Jaya)	(Jaya, 1 R-8 / Masuvi)	
(Guddu / Kayame)	Kayame				Shakthi	
(Kayame / Kavalagarru)	(Athikara / Jorrur)					
Kayame			Jaya			
(Kayame / Rajakayame)	Athikara J. B			Masuri	(Jaya / Shakthi)	
	P. T. B	(Kawalu- Kanna)				
Kayame	(J. B. / Athikara)			Jaya	Shakthi	

Chap.4 AGRICULTURAL LAND UTILIZATION AND PRODUCTIVITIES IN NARAVI

Dry land rice							
Acreage		Gross production	Seed (kg)	Productivity (Q/Acre)	Yield-seed ratio	Gross-Income per acre	Consumption
Planted	Harvested							
4.20	4.20	30	160	6.7	18.8	861.3	30
1.00	1.00						40
							30
							24
							68
							5
							22
							22
							30
							36
							50
							16
							30
							55
							25
							50
							26
							30
							140
							40
							160
							70
							70

Estimated yields of major varieties of paddy

Varieties	Yield (Q/acre)
Intan	18–25
Madhu	
MR–301, MR–272	20–25
Red Annupurna	
IR–20	
Vikram	
Sona	
Vani	25–28
IET–1444	
IET–2254	
Shakti	
Jaya	30–35

Source : *Package of practices for high yields*, Univ. of Agricultural Sciences and Dept. of Agriculture, Bangalore.

Table 4–4 Productivities of class-wise wet land

Category	Wet I	Wet II	Wet III
Average of 25 households ($\bar{x}$)	9.11	9.60	8.40
Standard deviation (a)	5.80	6.56	5.34
Coefficient value (c.v.)	63.67	68.33	63.57

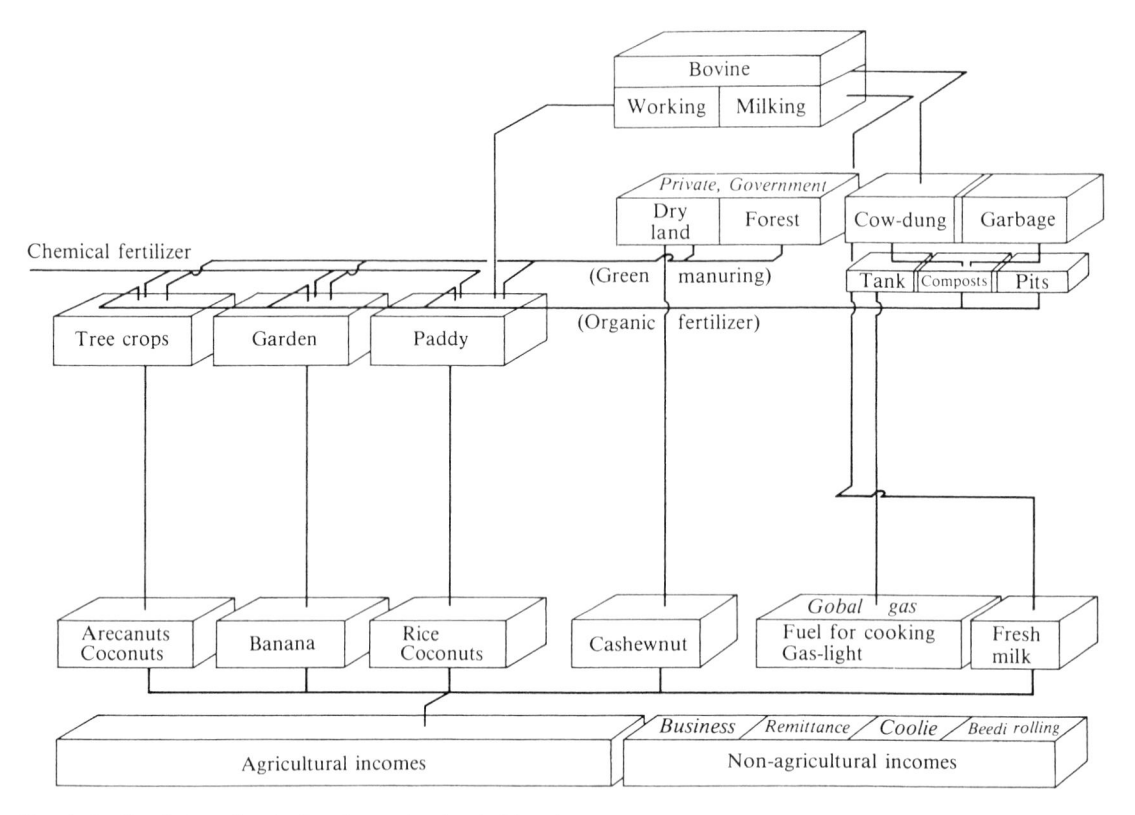

Fig. 4–3 Circulation of agricultural reproduction in Naravi

ied combinations of their income sources. Taking the typical case of intensification and diversification of the agricultural activities the householder No. 2–142 has recently ventured to keep poultry farm. He has got bank loan on the long term amounting to 20,000RS, part of which has been used for the repairment of pompset as well as for the newly introduced *gobal* gas system. He did his work successfully. Poultry farm and paddy field yielded fairly good revenues last year, it resulted in 35,000RS per annum. This sum is second largest among the 25 samples. Some of other households are doing non-agricultural work as business, or *beedi* (leaves of tobacco) rolling as a side job, some are supported by the remittance from a member of family, working away home, for example, Mangalore, Bombay and Gulf countries. Practically speaking full-time agricultural households add up to only 9 families (Table 4–6 and Fig. 4–3).

Among many items of agricultural management costs, main factor costs are the payment to labourer, the purchase of chemical fertilizer, agricultural machines and tools, and animal feeds. High-rise of agricultural factor costs have been preceded by the increasing employment opportunities for non-agricultural sectors.

Thanks to recent improvement of traffic condition, *beedi*-rolling has been introduced as a side job in this village. A lot of marginal workers were lured by a casual work on a comparatively high piece-rate bases. This is a one of the leading factors to push up wage level in the village. Because most of the tenants and the marginal workers, who worked under the big landowner was released as a excessive labourer after the enforcement of Karnataka Land Reform Act.

Government of Karnataka distributed a sizeable extent of land to these marginal workers including bonded labourers. While they cultivate the granted land, they are engaged in non-agricultural works as *beedi* rolling. Such marginal workers is very responsive to the varied rates of wages as well as labour conditions, and tend to select the best among many works as a remunerative

Chap.4 AGRICULTURAL LAND UTILIZATION AND PRODUCTIVITIES IN NARAVI

Table 4–5 Frequency distribution of productivities of paddy in the villages of the Western Ghats (Naravi, Kurubathur and Yadavarahalli)

Productivities (Q/acre)	Productivities (Q/acre) Naravi Wet I	Wet II	Wet III	Total	Kurubathur & Yadavarahalli Wet	Yields-seeds ratio Naravi Wet I	Wet II	Wet III	Total	kurubathur & Yadavarahalli Wet	
1											
2	1	1	1	3 ⎫							
3		3		3 ⎬ 15		1			1 ⎫ 1		
4	1			1 ⎩ (33.3%)					⎬		
5	3	3	2	8 ⎭					⎭		
6	2	3		5 ⎫			1		1 ⎫		
7	3			3 ⎪		1			1 ⎪		
8	1	1	1	3 ⎬ 21	2 ⎫ 6	1	1		2 ⎬ 8		
9	1	2	1	4 ⎪ (46.7%)	1 ⎬ (35.3%)	1	1		2 ⎪		
10	2	3	1	6 ⎭	3 ⎭			2	2 ⎭		
11					1 ⎫	1	1		2 ⎫		
12		1		1 ⎫ 1	2 ⎪	1	1	1	3 ⎪		
13				⎭ (2.2%)	⎬ 11				⎪		
					⎪ (41.2%)				⎪		
14					3 ⎪	1	3	1	5 ⎬ 15		
15					1 ⎭	2	3		5 ⎪		
16				⎫	1 ⎫	2	2	1	5 ⎪	1 ⎫	
17				⎪	⎪				⎭	⎪	
18	1	1		2 ⎬ 5	⎬ 4				⎫	⎪	
				⎪ (11.1%)	⎭ (23.5%)	1			1 ⎬ 7	⎬ 1	
19	1	1	1	3 ⎭					⎪	⎪	
20					3 ⎭	1			1 ⎭	⎭	
21				⎫					⎫	⎫	
22				⎪					⎪	⎪	
23	1	1		2 ⎬ 3					⎬ 6	⎬ 1	
24				⎪ (6.7%)					⎪	⎪	
25		1		1 ⎭		3	2	1	6 ⎭	1 ⎭	
26									⎫	1 ⎫	
27									⎪	⎪	
28									⎬ 1	⎬ 2	
29									⎪	⎪	
30							1		1 ⎭	1 ⎭	
31										⎫	
32										2 ⎬ 3	
33										1 ⎭	
34											
35											
36											
37											
38											
39											
40							1	1	2 ⎫	3 ⎫	
41									⎪	⎪	
42									⎪	⎪	
43									⎪	⎪	
44									⎪	⎪	
45									⎪	⎪	
46						1	1		2 ⎬ 5	1 ⎬ 7	
47									⎪	⎪	
48									⎪	2 ⎪	
49									⎪	⎪	
50							1		1 ⎭	1 ⎭	
61-70										1	
71-80						1	1		2		
81-90											
91-100										1	
100<										1	
Total	17	21	7	45	17	18	20	7	45	17	16*
Min.	2.4	2.7	10.0	2.4	8.3	3.0	6.7	6.7	3.0	16.7	16.0
Max.	23.3	25.0	19.0	25.0	20.7	75.0	75.0	40.0	75.0	133.3	90.60
x̄	9.11	9.60	8.40	9.23	13.46	20.17	22.07	10.19	21.06	47.11	41.73
a	5.80	6.56	5.34	5.99	4.05	16.75	17.52	10.93	16.14	28.09	17.78
a/x	63.67	68.33	63.57	64.90	30.00	83.04	76.61	60.12	76.64	59.63	42.56

* The sample of the house No. Y-16 was excluded from caliculation

Table 4-6 Economic performance of farm management

	Household number	1-1	4-2	4-3	4-172	3-4	2-149	1-51	2-86	1-74	2-166	2-140
Gross income (Rs)	Agriculture	500		5,000	4,200	3,000	10,000		5,000	5,000	15,000	6,000
	Cattle breeding		720		5,000							
	Land rent											
	Casual agri. labourer	1,500										
	Business						30,000					
	Working away from home	4,800				1,000						4,800
	Other			4,800				1,200*			500**	
	Total	6,800	720	9,800	9,200	4,000	40,000	1,200	5,000	5,000	15,500	10,800
Farm Expenditures (Rs)	Chemical fertilizer		330	400	1,000		1,300			150		1,500
	Organic fertilizer		500	400	1,000	1,500	1,500		2,000	800	2,000	2,500
	Insecticide, herbicide			100		200			200		200	100
	Seeds and young plants		140		320		300	115				300
	Animal feeds		1,000	250	3,600		5,500	720	500	500	500	1,200
	Fuel and electlicity			150			300					900
	Payments to labourer		805	2,000	2,000	1,600	1,200	M.S.	1,000	800	3,000	4,200
	Land rent											
	Costs for irrigation		50				1,500	9				60
	Transportation costs						140					200
	Purchase of small animals											
	Purchase of large animals		500				375					
	Purchase and repairment of farming tools and machines		75		200		1,500	45				500
	Land tax			40	15	122	150	20	22	22	70	75
	Other farm expenditures		100									
	Total farm expenditures		3,500	3,340	8,135	3,422	13,765	909	3,722	2,272	5,770	11,535
Living expenditures and loan	Monthly living expenditures of household		500	2,000	1,500	750	2,000	500	450	600	500	700
	Loan											
	Amount				3,000		5,000					8,500
	Interest rate (%)				10.5		16.0					11.0
	Purposes											
	1				Fertilizer		Payment of wages					Planting coconuts
	2				Payment of wages							
	3											

Ms refers to *Majuri* system (lafour exchange system)
*   Salary (Priest)
** Beedie rolling

Chap.4 AGRICULTURAL LAND UTILIZATION AND PRODUCTIVITIES IN NARAVI

2-5	2-6	1-71	2-138	2-7	2-38	3-190	1-18	2-8	4-9	1-10	2-142	2-87	1-16
8,000	8,000	10,000	20,000	10,000	10,000	5,000	15,000	15,000	30,000	40,000	35,000	16,000	10,000
								2,000					
720													
						4,000	2,500			10,000		3,000	
	12,376		20,000		21,600							27,400	
8,720	20,376	10,000	40,000	10,000	31,600	9,000	17,500	17,000	30,000	50,000	35,000	46,400	10,000
	500		600	2,000	1,200		1,500	300	2,500	700	1,000	2,400	500
1,000	600	2,000	1,000	1,500	1,000	3,000	500	1,000	500	3,000	2,000	3,000	2,000
	40		250	150	100	300	300	200		500	150	200	150
	50		1,000	100	250	500		150		600	200	300	200
1,500	1,000	1,100	350		6,000	500	2,000	1,500	1,800	3,000	3,000	3,000	2,500
	500		500		250		1,000			3,000	300		
1,800	2,000	2,000	6,000	2,100	5,000	3,000	5,000	5,000	6,000	6,000	5,000	8,000	3,000
			200	60							68		
			500	150		200			600		500	200	500
											120		
									5,000				
									17,300		5,000	2,100	
45	60	25	125	65	55	24	60	48	130	150	50	85	62
								200					
4,345	4,750	5,125	10,525	6,125	13,855	7,524	10,360	8,398	33,830	16,950	17,388	19,285	8,912
500	2,000	300	600	700	400	600	500	600	3,000	2,000	700	800	800
	4,000	700	5,000			10,040	10,000	5,000	4,000		20,000	10,000	1,500
	18.1	12.1	14.0			9.5	12.0	17.0	10.0		12.5	13.25	18.1
						11.5					(9.5)		
	Payment of wages	Payment of wages	Seeds			Coconuts	Pumpset repair-ment of pumpset	Aquisition of land	Tiller		Poultry	Pumpset	Payment of wages
	Pumpset					Cashew-nuts					Pumpset	Tiller	
						Irrigation					Gobal gass		

| 2-5 | 2-6 | 1-71 | 2-138 | 2-7 | 2-38 | 3-190 | 1-18 | 2-8 | 4-9 | 1-10 | 2-142 | 2-87 | 1-16 |

Table 4–7 Plantation and horticultural crops in Naravi

House No.	Area planted (Acre)	Crop	Number of plant	Yield	Selling price (Rs/kg)	Input of fertilizers (kg) Organics	Input of fertilizers (kg) Chemicals
4–2	0.10*	Cashewnut	10	15 kg			
	0.20	Coconut	25	0			125
		Arecanut	50	75 nuts			150
2–86	0.20*	Cashewnut	30	50 kg	7.0		
3–140	0.20	Cashewnut	18	100 kg	7.0		
		Coconut	100	500 nuts			5000
		Arecanut	10	500 nuts			
2–6	0.20	Coconut	60	800 nuts			21,000
		Arecanut	150	400 kg	16.0		
2–138	2.00*	Cashewnut	70	500 kg	3.0		700
	0.20	Coconut	25	2,000 nuts	@1.5		250
		Arecanut	300	20,000 nuts			3,200
2–38	0.10	Coconut	40	500 nuts			2,500
		Arecanut	100	10,000 nuts			1,000
3–190	3.00*	Cashewnut	230	200 kg	7.0		6,900
	5.00	Coconut	300	500 nuts			30,000
		Pepper	60	0			500
1–18	0.20	Coconut	50	500 nuts		100	1,500
		Arecanut	300	3,000 nuts			
2–8	1.10*	Cashewnut	125	70 kg	5.0		
	0.20	Coconut	70	500 nuts			
		Arecanut	150	70 kg			
4–9	side crop	Coconut	100	1,000 nuts	@1.5	25	500
	3.00	Arecanut	200	2,000 kg	10.0	200	2,000
1–10	2.00	Coconut	400	4,000 nuts	@1.5		70,000
		Arecanut	500	200 kg	12.0		50,000
2–142	5.00*	Cashewnut	325	200 kg	7	500	5,000
	1.00	Coconut	100	800 nuts			
2–87	2.00	Coconut	150	1,000		150	4,500
1–16	0.20	Coconut	30	400			700
		Arecanut	150	3,000			

* Dry land

side line. Thus big farmers, who depend on hired labourers, become incited by emulation and outbid each other. Pay-off balance of agricultural managements have been seriously aggravated greatly by keen competition for decreasing labourers who tend to demand higher wages and advantageous labour conditions.

In response to aggravating conditions of management, many of agriculturists have tried to make better use of exiting land.

3. Horticultural Crops

They planted tree crops like cashewnuts, coconuts and arecanuts. Some coconuts trees have been traditionally planted as a side crop on a levee and footpath between paddy field, as these tree crops require a plenty of waters. But nowadays these tree crop become planted on a horticultural gardens on a full scale. Table 4–7 show the distribution of tree crops and its economic pay-off balances among the sample surveyed farmers. Major tree crops are concentrated on cashewnuts, arecanuts and coconuts. Only one farmers (3–190) cultivates pepper on 5 acres of land along with coconuts. Recent growing demand for home and abroad has accelarated the cultivation of cashewnuts. Unit price (Rs/kg) is not as much high as arecanuts.

But from the view point of intensification of land use as well as one of possible way to make better use of unirrigated cropland, cashewnut is among the most advantageous commercial tree crops. These tree crops can be planted on the dry land. In fact there are 9 farmers in the samples. It is often pointed out that cashewnuts grow even on a gravelly soils and the adequate soil moistures are continuously needed for a successfull cashew plantation. But here in Naravi village, substantially, no careful attention are necessary to cultivation, irrigation or manuring except pruning of the dead branches and clearing of under growth. The trees of cashewnuts yield some crops, but not adequate to pay off. This is mainly because these trees crops in the sample have recently planted. The first bearing is normally secured in three years after planting. A crop reach in its peak after eight years. Arecanuts is not profitable, however, investment for irrigation facilities amounts to a sizeable degree.

This is principal deterrent factor for further expansion of arecanuts.

In addition cultivation of arecanuts require a continued careful attention to diseases as fruit rot, stem bleeding, root rot, leaf bright and so on. However these horticultural plantation could prove as profitable as any other revenue-raising scheme like poultry and swine industry in the near future to come.

CHAPTER 5

KARNATAKA LAND REFORM ACT AND ITS IMPACTS ON THE SOCIO-ECONOMIC CONDITIONS OF NARAVI

1. Preface

The land reforms in rural India have been long overdue and stringent problem to be solved. In an attempt to vitalize Indian economy mainly based on primary sector, Central Government has taken a series of legislative counter-measures necessary for the reorganization of land tenure system.

The Central Government stressed that land reform is to provide fastidious socio-economic base and institutional framework for bringing about not only an increased productivities but also the modernization of rural India.

Land reform has been considered to be as a precondition to the processes of rapid economic growth. Thus special emphasis has been placed on the reorganization of land tenure system, aiming at radical transformation of agrarian structure in the due course of developmental planning.

In accordance with these legislative steps by the Central Government, Government of Karnataka has been hammering out a series of effective land reform aids.

The Karnataka Land Reform Act of 1961, as amended up to 7th March, 1980 opened the way to the modernized land tenure system of present day. The Karnataka Land Reforms Act of 1974 was intended to expediate overdue land reforms as well as to close loopholes in the processes of enforcement, which was keenly felt among the policy makers. In this respect the Act of 1974 was highly esteemed by the authorities and the parties concerned except the landed interests.

2. Land Reforms in Naravi; an Overview

The sample village, Naravi remained backward and undeveloped till the inception of Community Project in 1952. Since then transportation and communication system have much improved and various development programmes have ameliorated the socio-economic conditions of the villages to a great extent. Among these rural development policies the most important factor influencing economic life in the sample village is considered to be land reforms policies by the Government of Karnataka state. Mysore Land Act worked out in 1962 to fix the rent at twenty-five of the yields as for the wet land.

However the Act proved to be ineffective and remained an empty statutes. There has been an increasing controversy over the outdated system of land tenure. The enforcement of Karnataka Land Reforms Act of 1974 has greatly transformed rural community as well as traditional landlordism all over the state. The sample village, Naravi has been much affected by the land reforms. Socio-economic conditions of the village have dramatically changed in recent years. All the things were disturbed and deranged completely. Radical changes in landholdings have resulted in the upset of decaying landlordism.

At the same time the narrowly-balanced socio-political relationships between landlords and tenants have been overturned by the growing influences of Christianity in Naravi. On the other hand, communism movement, acclaiming radical land reforms has deeply rooted in Kerala state. Their overwhelming rampant political influences propagated itself all over the adjacent regions including South Kanara, and come to be keenly felt among most of the weaker sections of this village.

These factors put spurs to speed up a radical land reform in Naravi village. The EX, Chief Minister of Karnataka, Mr. F. Devaraje has perfectly carried out a series of radical land reforms acts to realize social justices as well as an increased agricultural production at the same time in accordance to these uprising powers.

As the result of these continued efforts, *Dani* system was completely wiped out. It was the real scene in the 1970's as though the world had turned upside down.

In this chapter, auther tried to make an objective appraisal of the various socio-economic impacts caused by the Karnataka Land Reforms Act of 1974 and trace discernible recent trends in the rural country.

3. Landlordism in Naravi before the Land Reforms Act

Before the enforcement of Karnataka Land Reforms Act, *Dani-Vokkalu* system had prevailed widely over South Kanara region. *Dani* means literally landlords and *Vokkalu* refers to their tenants and bonded labours. This landlordism had long been observed in this locality. *Dani-Vokkalu* system can be divided into two types of agricultural management according to its labour organization.

The one type of *Dani-Vokkalu* system was based on a band of labour gang, where agricultural workers were organized into the several units of labourers group and were under close surveillance of a landowner. Landowner himself cultivated the land of his own with them. These agricultural labourers were composed of many a landless peoples who could not afford to have homes and lands of their own. A few of landowner often constructed shabby labourers quarters for them and provided necessary goods of various kinds for their daily use.

The other type of *Dani-Vokkalu* system was heavily dependent on the tenants and labourers. In this system, landlords leased out most of their lands to many tenants competing each other to get favorable tenant rights among them. Tenant cultivated the leased-in land on his responsibility and accounts. There were some cases that the cultivator of those leased-in lands had a few acres of lands.

In this sense, this type of cultivator could be identified as owner-cultivator at the same time tenant. Some of them sublet his leased-in land to other tenants.

The rent was paid to landlord in the form of rice, which was sometimes husked. Because price of rice remained stable in the long term and considered to be as a standard unit in the barter trade.

As a matter of fact village economy had not been fully fledged in those days. The stable value of rice was equated with that of other commodities of various kinds. This means the fact that rice had played a part of critical importance as an equalizing agent in various commodities, which had different exchange values. As for the payment of rent, the fixed rent was widely practiced for long time with the equal share of 50% for the landlord to 50% for the tenant of the total yields. The rent to be paid to the landlord had been fixed up to 1974.

But amount of rent was in fact dependent on the actual productivity level of each leased lands as well as socio-political balance between landlords and tenants. There can be seen some cases that two-third of the total yields should be paid to landlord. There is also a few cases that the rent is expressed in the absolute term, *viz*, as in 3 quintals and 20kg of rice per acre. It depends on actual situations. It is very interesting to note the case that the first paddy i.e. *yenel* belongs to landlord and second crop, *suggi* to the tenant. This is a sort of crop-sharing. In any cases terms of payment of the rent depends on actually on various situations.

As also in *Dani* system based on leasing out to tenants, the landlords not only constructed houses for the tenants and their agricultural labourers but also extended their helping hands to tenants in which included financial assistances on a marriage function as well as sizable supports in kinds on various occasions.

In addition to these beneficial assistances, landlord had reduced the land rent in the case of successive crop failures for the conveniences of tenants.

The former type of *Dani* system was based on the band of agricultural labourers including the bonded labourers, and managed by the big sized owner-cultivator in nature.

On the contrary, the latter type could be identified as absenteeism in spite of the fact that most of them really live in the village with their family. Because the land-owners were mostly detached from the actual cultivation of land of their own.

The almost all of their lands were leased out usually to landless tenants. These tenant cultivate the land for his own account and responsibilities. Managerial accountability lies on the tenants side. All the landlords do is to receive the fixed rent from the tenants.

Seeing from the hierarchical relationship between landlord and tenants, they form a stable micro-cosmos in the rural community, where landlord takes on patriarchial leaderships over all the members to survive in the subsistence economy.

As far as tenants and agricultural labourers accept the proposed arrangement as well as various sort of regulations by the landlord, peaceful coexistence could be safely maintained in the patripotestal regime. This assignment is not always a paying concern.

In this sense, *Dani* system shall be justified by the fact that it had played a vital part in rural community living on the subsistense level. However, from the view of modern concepts relating to land tenureship in the capitalist societies, *Dani* system is outdated in many respects.

On the other hand, in so far as widening gap between a handful of the affluent class and the poor mass in rural area has been keenly felt among policy-makers or opinion leaders, and its main cause is attributable to the uneven distribution of landholdings as well as the outdated land tenureships.

It was very reasonable that the radical land reforms should be taken up as a first policy measures against vi-

Table 5–1 Categories of agricultural land classified by the type and method of irrigation and major crops

Class		Type and method of Irrigation	Major crops
A		Assured irrigation from Government canal and tanks	Two crops of paddy or one crop of sugarcane in a year
B	I	Assured irrigation from Government canals and tanks	One crop of paddy in a year
	II	Irrigated by lift irrigation projects constructed and maintained by the State Government	Two crops of paddy or one crop of sugarcane
C	I	Irrigated from any Government sources including lift irrigation project constructed and maintained by State Government except Class A and Class C	
	II	Rain water	Paddy or areca
	III	Irrigated by lifting water from Government canal or tank where the pumping installation or other device for lifting water is provided by the land owner	Light irrigated crops other than paddy and sugarcane
D		Dry land but not having any irrigation facilities from Government sources	Paddy or garden crops not coming under Class A, Class B or Class C

Table 5–2 Formula for determining equivalent extent of land of different classes

Acre(Unit)

Class	Soil classification value	
	Above 8 annas *	Below 8 annas
A	1.0 (1.000)	1.3 (0.769)
B	1.5 (0.667)	2.0 (0.500)
C	2.5 (0.400)	3.0 (0.333)
D	5.4 (0.185)	

* 8 annas is equal to 50 paise

Table 5–3 Standard acre

Class	Standard acre
Wet I	2 acre
Wet II	4 acre
Wet III	4 acre
Dry land	4 acre
Garden	1 acre

cious circle of poverty in rural areas. In fact, Government of Karnataka State ventured to enact all the necessary amendments for the radical land reforms in 1974.

4. Legislative Aspects of Radical Land Reforms

The Karnataka Land Reforms Act of 1962 has been amended ninteen times up to 1980. Radical aspects of the act are clearly reflected on the several sections and sub-sections. Whatsoever these sections may be substituted or amended even added, it is an undeniable fact that the Government of Karnataka State has been fully determined to carry out the drastic reforms of land tenure system with a purposeful intention to increase an agricultural production as well as to realize social justice through silent euthanasia of the wealthy landed classes. We shall examine these processes minutely, with a special reference to the relevant sections in the act and also to the results of village study.

The salient features revealed in Karnataka Land Reforms Act could be reduced into the following several points.

1) To abolish landlordism there was prohibited leases of agricultural land, as expressed in section 5 of chap. 2, that no tenancy shall be created or continued in respect of any land nor shall any land be leased for any period whatsoever.

2) The same thing has been said concerning the difinition of "cultivation personally", as defined to the effect that to cultivate personally refers to cultivate land on his own account, by the laour of cultivator himself or family member, or by servant on wages payable in cash or kind. However, the most important point is the prohibition of cultivating land on the crop sharing basis, under the personal supervision of oneself or by member of one's family.

Those two sections gave a decision blow to the landlordism as *Dani-Vokkalu* system. According these prohibition clauses, landowner who leased out his land to tenant shall be punished on the ground that landowner leased out his land to tenant and he did not cultivate his land his own account. Even if he cultivated it by himself or any member of family or hired labourer, he should not get them worked under close surveillance of himself on the crop-sharing basis. *Dani* system were destined to disappear sooner or later.

3) The Ceiling area was clearly difined in a practicable way. This clause led most of big landowner to silent euthanasia except those who cultivate plantation crops in which include cardamon, cocoa, coffee, pepper, rubber and tea.

The radical land reforms worked themselves out by the postulation of ceiling on land. As postulated in section. 63. Chapter IV, no person shall be entitled to

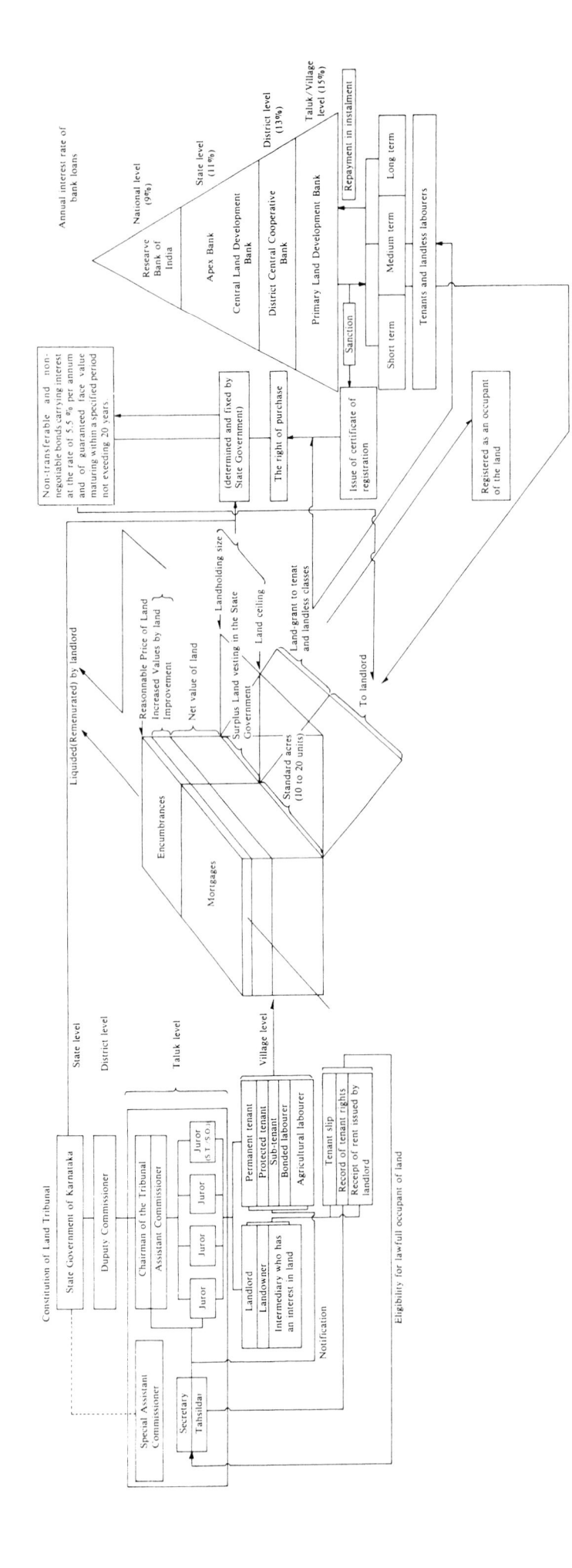

Fig. 5-1 Legislative framework and Processes of land reforms in Karnataka State

Table 5–4 Grant of surplus land to Scheduled Castes and Scheduled Tribes

Priority	Category of applicant	Extent of land-grant
1	Dispossessed tenants who are not registered as tenant	Not exceeding one unit each
2	Displaced tenants having no land	do
3	Landless agricultural labourers	do
4	Landless persons and ex-military personnel whose gross annual income does not exceed Rs.2,000	do
5	Released bonded labourers	do
6	Other persons residing in villages of the same *Panchayat* and whose gross income does not exceed Rs. 2,000	Not more than the extent required to make up to one unit

hold land in excess of the ceiling area. The ceiling area means upper limit of the extent of agricultural landholding. In other word, no person shall not be entitled to hold land in excess of the ceiling area. In calculating ceiling area, all the lands is to be converted into the standardized unit of land (Table 5–1, 5–2 and 5–3).

Criteria to determine the ceiling on land is as follows; a person who is not a member of family or who has no family or for a family should not hold land more than 10 units. Provided that in the case of a family with more than 5 members, ceiling area shall be 10 units plus additional 2 units for every member more than 5, however it should not exceed over 20 units in total. Accoding to section 65, surplus land, more than the preserved ceiling area, shall be surrendered to the Government of Karnataka State.

In the case that land has been converted into any other class as a result of irrigation from source constructed by State Government, and consequently turned out to exceed the ceiling area, the land in excess shall be deemed to be the surplus land.

Those who hold such land should furnish a declaration to *Tahsiladar* the details and particulars. All these surplus land is to be surrendered to the State Government.

In order to confer ownership on tenants, the Land Reforms Act has hammered out a series of stipulations with effect from 1-3-1974 on the vesting of land in the State Government.

This is the main objective that State Government aims at. Thereon Section 44 of Chap III states that all rights title and land interests of all lands shall be transfered to and absolutely vest in State Government free from all encumbrance on and from the date of vesting and all the land once freezed is to be distributed among the permanent tenant, protected tenant and subtenants.

These tenants are lawfully entitled to be registered as an occupant (registered occupant). "Permanent tenant" refers to a tenant cultivate on his own account and protected tenant to a tenant who has held continuously and cultivate land on his own account for more than 12 years.

They are entitled to be registered as a lawfull occupants and obtain land for counter value. In other word, all the land exceeding over ceiling stand transfered and vest in State Government and come into their possession on an onerous basis.

Seeing from the view point of landowners, and landlords, they are entitled to claim an amount determined with reference to the net annual income from the land for the extinguishment of their rights in the land (section 47).

This section is also applied to the following case; namely, landowner who has taken possession of any land by evicting a tenant in order to cultivate it on his own account or use it for non-agricultural purposes and thereafter he failed to cultivate it within one year or ceased to cultivate within three years, the land shall stand transfered to and vest in the state government free from all encumbrances (section 22).

Tahsildar shall fix the reasonable price of the land, viz, amount of payment and determine any mortgage or other encumbrances lawfully on the land, which are charged on the person who has created the mortgages of encumbrances,

At the same time *Tahsildar* shall determine an estimated value of land improvement by the tenant and amount payable. This is compensated by the landowner. The reasonable price shall be fixed according to the following procedures; either landlord or tenant apply in writing to the *Tahsildar* for determining the reasonable price. Then *Tahsildar* shall fix it, considering average prices of similar lands in the locality during ten years into account.

The tenant shall deposit with *Tahsildar* the amount of land. *Tahsildar* shall issue a certificate declaring him to be the purchaser of the land.

In the next place. We examine minutely about the mode of payment of amount payable to any person prescribed under section 47. Because it is one of the most important points to make a critical appraisal of Land Reforms Act of Karnataka and its socio-economic impact on the village community in Naravi.

Firstly, the landlord shall be paid in cash at one

time for the extinguishment of their rights in land, if the amount does not exceed 2,000 Rs. But if the amount exceeds over, the amount shall be paid up to 2,000 Rs. The remaining amount is to be paid in the form of bond. This bond is non-transferable and non-negotiable in nature, carrying interest at the rate of 5.5% per annum.

In addition the holder, viz, landowner could encash the guaranteed face value after maturing within a specified period not exceeding 20 years. Provided that all the amount shall be paid in lumpsum in the cases that a minor, woman who has never married, a person whose annual income from all sources is not more than 4,800 Rs, the phisically and mentally handicaped and widow.

Every tenants who is registered as an occupant is titled to get term loan from the State Land Development Bank or a Credit Agency with the annual interest of 5.5% for the payment of premium. The amount of premium shall be payable to the State Government. The period of payment should not exceed over twenty years, provided that where payment is in instalment, 2,000 Rs share be paid as the first instalment and balance in an equated annual instalments.

Finally we come to the sections dealing with the disposal of surplus land in the Land Reforms Act of Karnataka of 1961.

Surplus land vesting in the State Government is granted to the categorized weaker sections as shown in Table 5–4. The surplus land vesting in the State Government are composed of the following items.

1) The land of which tenant was not entitled to be "registered occupant" before the date of vesting.
(Section 45–3)

2) The land which was leased contrary to the Act.
(Section 58)

3) The land which was acquired by any one who has an assured annual income more than 12,000 Rs from sources other than agricultural land.
(Section 79A)

4) The land held by a person other than cultivating land on his own account or institution, trust, company, association, cooperative society except cooperative farm whose land are not used solely for their own purposes. (Section 79–B and 63–7).

Those lands are granted to the categorized peoples. As clearly prescribed in the section 26 of Karnataka Land Reforms Rules of 1974, the conditions and guidelines for grant of surplus land are expressly in the text as follows;

1) The grantee shall cultivate the land on his own account.

2) The land shall not be appropriated for any purposes other than agriculture.

3) The grantee shall plant not less than five fruit-bearing trees per acres and thereafter maintain the said tree before the completion of the first agricultural season.

The 50 per cent of the land vesting in the State Government are granted to Scheduled Castes and Scheduled Tribes specified in the Constitution (Scheduled Castes and Scheduled Tribes Order; 1950). The remaining 50 per cent of the land shall be reserved and then distributed to the peoples belonging to weaker section other than S.C and S.C mentioned above at the same conditions. (Tab 4) However as a rule, they can obtain these land for counter value as stipulated in the section 78.

It states to the effect that the grantee shall have the option to deposit with the Tribunal the amount of price of the granted land in either in a lumpsum or in an annual instalments not exceeding 20 years. Annual interest rate is 5.5%. Purchase price is equal to 15 times the net incomes in the case of A, B and C classes of lands. The deputy commissioner makes final decision on the disposal of the surplus land.

5. Land Reforms in Naravi and its Critical Appraisal

The land reform reached at its peak in the 1974 and thereafter socio-economic condition of Naravi village has tremendously changed by the land reforms. *Dani* system has been completely wiped out in this village.

In those days, *Dani* had extended their helping hands the weaker sections of landless poor villagers.

Dani began to lease out their land and constructed shabby houses for tenants called *Vokkalu*. But in due course most of them reduced to the bonded labourer. These labourers were getting out of the clutch of declining *Dani* system by the Radical Land Reforms Act of 1974. They found their way and rose from servile labourers to owner-cultivators. As have pointed out, the land reforms give rise to not only modernized system of land tenure, but also levelling out of land-holding.

We shall examine these processes by taking up typical case in Naravi village. The informant (Household No. 2–87) was one of the most biggest landowners before the land reforms and belongs to the dominant caste, Jain. Before the enforcement of Karnataka Land Reforms Act of 1974. He had 284 acres of agricultural land in total. Out of which 169 acres (59.6%) of land were in Naravi. The remaining land were distributed among other three villages. He has land leased out to 25 tenants in all. Out of which 21 tenants (84%) were in Naravi village. The proportion of the total land leased to the total wet land in Naravi are calculated to

Table 5–5 Landholding and tenants of household No.2-87 before the land reforms

Location		Tenant		Extent of land	
		Number	%	Acre	%
Inside the Village (Naravi)		21	84.0	169	59.5
Outside the village	Idhu	2	8.0	50	17.6
	Nuraluvettu	2	8.0	40	14.1
	Marodi			25	8.8
Total		25	100.0	284	100.0

Table 5–6 Landholding and incomes of the household No.2-87 in 1982

Income sources	Income
Agriculture Total extent of land : 20.20 acres Wet II (3.20), Wet III (9.00) Garden (2.00), Dry land (6.00)	Rs 16,000
Business (general shop)*	Rs 3,000
Remittance (working away)	Rs 27,400

* Opened in 1974

Table 5–7 Details of hired agricultural labourers of the household No. 2-87 in 1982

Age	Sex	Place of birth	Payment	Since when
40	M	Naravi	Rs 8 per day 3 meals and 2 coffees	1970 (12)
35	F	Naravi	1 meal, 2 kg of rice and 1 coffee	1964 (18)
45	M	Aladangadi	Rs 8 per day 1 meal and 1 coffee	1979 (3)
42	M	Aladangadi	Rs 8 per day 1 meal and 1 coffee	1980 (2)
38	F	Kela	1 meal, 2 kg of rice and 1 coffee	1977 (5)
45	F	Reuyada Karkal *Taluk*	1 meal, 2 kg of rice and 1 coffee	1977 (5)

be 21.9%. It means that he occupied substantially one forth wet land in Naravi village (Table 5–5). With these economic backgrounds he had exerted powerful influences over villagers and his tenants as a *patel* (village-head) for long time. But as the land reforms have been increasingly in full swing in 1970's, the strongfold of *Dani* system was broken into pieces. The threat of radical land reforms has been keenly felt among the landlords in Naravi and in fact has shaken up the entire village. In due course, he has lost 263.5 acres o and. The total extent of 263.5 acres of land were identified as surplus land and transfered to a State Government. All the land did not become vested in the State Government. As some of debatable land are still now pending in Land Tribunal at Belthangady, he has been summoned so often to the court. It is time-consuming matter for him to go and back. He has now only 20.20 acres out of which wet I and II are 12.20 acres, garden 2.00 acres and dry land 6.00 acres (Table 5–6). He gave up agricultural management in despair and so reduced it to the minimum size. This could be justified by the following facts. He hires 6 permanent labourers out of whom 2 workers have been working on his farm for more than twelve years. Other workers have been employed in these five years. Wages are paid both in cash with the rate of 8 Rs per day for men and in kinds, viz., 2kg of rice and additional meal and coffee, respectively. These payment adds up to 8,000 Rs per year (Table 5–7). The factor costs of this kind incur considerable expenses. Unfavorable pay-off balance of agricultural management is mainly due to the payment to labourers. He is now trying to hire in more seasonal labourers instead of permanent labourers. Factor cost of agricultural labour is not determined by the av-

erage productivity, but the non-agricultural conditions.

As a result of the land reforms, the bonded labourers were released from the clutch of *Dani* system which was organized by landlords, most of them have flowed out of agricultures sectors and changed their occupations.

They have tended to get more favorable employment among various sectors. The relative shortage of agricultural labour forces caused by the continuing outflows of labourers and increasing number of remunerative employment have brought about a sizable raise in pay on the whole. Consequently householder of 2–87 tries to reduce agricultural management to the self-supporting size.

The second reason for it is closely connected with an availability of money to invest. As has closely examined, the landowner shall be paid the amount up to 2,000 Rs for the extinguishment of rights in land exceeding the prescribed ceiling area, which stands in vesting in the State Govern-ment.

But the remaining amount of money is to be paid in non-transferable and non-negotiable bonds, carrying annual interest at the rate of 5.5%. The holder of this bond can not encash the granteed face value until maturing within a specified period not exceeding twenty years. Therefore the holder of this bond can not but to wait maturing of the specified period. During the specified period, he is not in a position to make full use of the bond in some ways. It is very restrictive in nature.

To make matter worse, the "reasonable price" of land means one tenth in value compared to the current price of the said land. Householder No. 2–87 has lost almost his land except the existing land of 20.20 acres. The percentage of the total land he had is only 7.2. It is less than one tenth.

In addition he received a non-transferable and non-negotiable bond which is equivalent to one tenth of the current value of land. This gave him a decisive blow and discouraged his vigorous incentives to invest for the improvement of agricultural productivities. There is few alternatives but to reduce agricultural management to the minimum or to cultivate his decreased land in an extensive way.

Furthermore the annual interest at the rate of 5.5% of the bond tends to be offset by the constantly increasing price level under the current inflationary circumstances.

At the time when the radical land reforms was in full swing, householder of No 2–87 ventured into business and opened general shop in 1974, forseeing oncoming impacts that he should much affected by the radical land reforms, the income from business amount only to Rs 3,000, with the proportion of 6.5% to the total income, however, the newly established shop shall yield more than ever before (Table 5–7). At leaset it is appropriate to say that this line is comparatively more profitable and promising than agricultural sector. But at present, this household depends heavily on the remittance. The amount get to nearly 60% of the total income. The former landowners can not live solely on agriculture. The significance of agriculture has much decreased in recent years.

Generally speaking, the former ruling class tends to be shabby-genteel even reduced in circumstances. But actual scene in this village shows the dispossessed landowner class can make nothing out of nothing and landless people made something out of nothing.

As for as we observed, most of land owners seem to come home to their home that "every flow has its webb". They are now trying to adapt themselves to the completely renewed milieu. For them, the land reforms means a process of euthanasia of landlord.

Turning our eyes to another aspect of land reforms and its impact on tenants as well as weaker sections, it comes out to be obvious that most of them have enjoyed the fruits of the land reforms. Land grant is among the most gainful effects.

The landless villager became owner-cultivator whatever the landholding size may be. Considerable extent of land came into their possession. They find themselves to be self-reliant, released from the burden of crop-sharing system and dispressed status as well. This led to an increased productivity of agriculture to some extent. However, this seems to have been counterbalanced by the decreased productivity and extensification of agricultural management on the whole.

As regards the average landholding size, either the middle class ranging from 5 to 10 acres and the marginal small class below 5 acres have remarkably increased in recent years. In this respect, one of main objectives of land reforms appears to be successfully realized.

Bountiful land grant was assured by the legislative intentions in the Land Reforms Act of 1974, Sections from 26 to 26AA. Generous distribution of surplus land vesting in the state Government could be justified by the following preconditions through successive intensification of land-use.

(1) to realize an increased productivities both in physical and monetary terms.

(2) to modernize traditional land tenure system as well as a release of bonded labourers from the clutch of landlord, provided that social justice could be realized as much as possible in many respects.

(3) to aim at levelling up socio-economic well-beings by the integrated policy mix in which the land reforms policy is included, lest the social system should fall into a deranged state.

Among these preconditions, the Land Reforms Act

Chap.5 KARNATAKA LAND REFORM ACT AND ITS IMPACTS ON THE SOCIO-ECONOMIC CONDITIONS OF NARAVI

Table 5–8 Long term trends in extent of agricultural land in Naravi

Year	Wet	Assessment	Garden	Assessment	Dry land	Assessment	Total	Assessment
1903	756.18	2086.1	70.64	196.15	50.93	38	877.75	
1915	756.80		75.36		175.10		1007.26	2374.8
1927	756.94	2088.2–0	78.22	207.7–0	420.80		1255.96	2474.10
1935	761.17	2363.13	82.15	245.2–0	483.70	200.7–0	1327.02	2089.6–0
1936	761.17	2363.13	82.24	245.5–0	489.37	202.10–0	1332.78	2811.12–0
1937	763.04	2373.10	82.52	246.6–0	496.66	205.5–0	1342.22	2825.5–0
1938–39	763.04	2373.10	82.52	246.6–0	500.31	206.11–0	1345.87	2826.11–0
1939–40	766.22	2384.9	83.74	249.4–0	500.42	205.11–0	1350.38	2839.8–0
1940–41	766.22	2384.9	83.74	249.4–0	499.41	205.4–0	1349.37	2839.1–0
1943	766.22		83.74		499.41		1349.37	2839.1–0
1944	767.31		83.74		499.24		1350.29	2840.4–0
1945	769.46		83.74		499.24		1352.44	2841.2–0
1946	767.46		83.74		506.34		1357.54	2844.3–0
1947	767.46		83.74		515.41		1366.61	2847.10–0
1948	767.46		84.68		517.69		1369.98	2850.12–0
1949	769.00		84.68		518.00		1371.68	2853.3–0
1950	769.00		85.68		520.45		1375.13	2856.6–0
1951	769.00		86.15		520.29		1375.44	2857.6–0
1952	769.05		86.15		521.12		1376.32	2857.13–0
1953	770.46		86.15		547.30		1403.91	2872.12–0
1954	770.51		86.15		556.80		1413.46	2876.9–0
1955	770.51		86.15		561.15		1417.81	2878.3 0
1956	771.13		86.21		586.87		1444.21	2888.12
1957	771.13		86.37		658.16		1515.66	2917.11–0
1958	771.13		86.37		678.27		1535.77	2925.13–0
1959	771.13		86.37		695.42		1552.92	2932.7–0
1960	771.51		86.37		721.96		1579.84	2943.0
1961	771.51	2396.93	86.37	256.0–0	735.43	295.12	1593.31	2948.1–0
1962	771.51		86.37		743.75		1601.63	2951.3–0
1963	771.51		86.37		782.62		1640.50	2965.13–0
1964	771.51		86.37		814.05		1671.93	2977.12
1965	771.49		86.37		813.81		1671.67	2977.13
1966	771.49		86.37		819.25		1677.11	2979.14
1967	771.49		86.37		819.25		1677.11	2979.14
1968	771.49		86.37		825.54		1683.40	2982.4–0
1969	771.49		86.37		840.36		1698.22	2987.13
1970	771.49	2396.15	86.37	256.00	840.36	334.14	1698.22	2986.29
1982	786.19		86.37		1869.87		2742.43	

Table 5–9 Recent changes in land use in Naravi

Year	Paddy field					Garden	Dry land
	Wet I	Wet II	Wet III	Others	Total		
1961	330.06	131.56	309.89	—	771.51	86.37	735.43
1970	—	—	—	—	771.49	86.37	840.36
1982	324.58	127.42	305.60	28.6	786.19	86.37	1,869.87
change (1961–82)	5.48	4.15	4.29		14.68	0	1134.44
Per cent					1.9	0	254.3

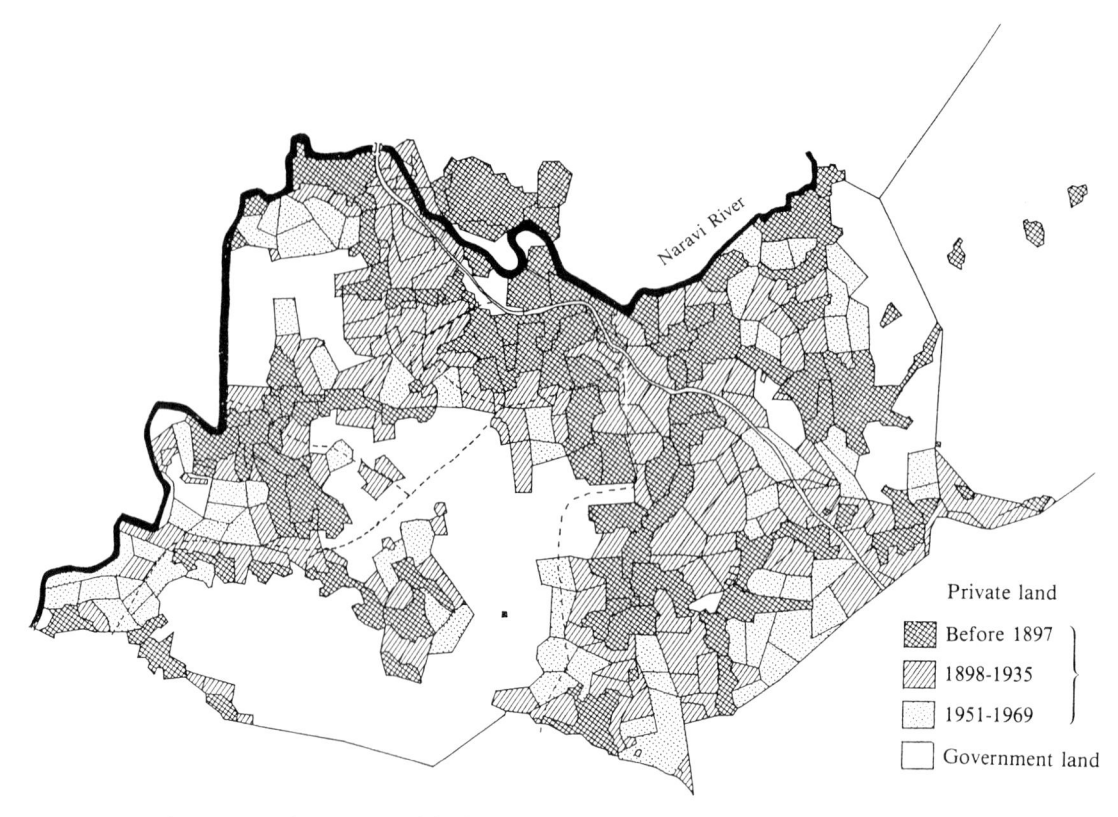

Fig. 5–2 Spatial expansion of private owned land

of Karnataka has been based only one precondition of (2). In other words, the radical redistribution of land alone has proceeded on exclusively and resulted in an unnecessary confusion and economic losses except modernization of traditional land tenure system.

In this sense, the land reforms were destined to be abortive from its very beginning. But it must be emphatically stressed that the Land Reform Act has opened the way to new era in the agrarian history of India and one of the major break-throughs in a time-honoured land tenure system. Everything has to be begun again.

It is appropriate to say that policy makers of the State Government have just begun with the redistribution of land. It is surely a new deal worthy of mentioning. New deal was promptly initiated by EX. Chief Minister of Karnataka, Mr. D. Devaraje Urs and his associates. Majority of villagers have pointed out that the main purpose of this land reforms act was nothing more than to make sure their seats in the Government.

Whatever their political ambition may be, agrarian revolution is unmistakably gearing up not only for improved agricultural production with new land tenure system, but for betterment of living conditions in rural India.

Remarkable increase of dry land validiates these discernible trends. Among the recent trends in various land use, dry land alone has shown a dramatical increase in acreage in this village.

Taking up the period from 1961 to 1982, there can be seen a phenomenal increase of dry land from 735 acres in 1961 suddenly up to 1,869 acres in 1982, with proportion of 253%. Meanwhile paddy has increased only 14.68 acres in extent and garden remained almost the same in these twenty years (Table 5–8 and 5–9). Judging from Table 5–8 and Fig. 5–2 showing long term trends in agricultural land use in Naravi village that spans ninety years from 1903 to 1982, we can trace the following salient trends revealed in changing land use; 1) As regards the land use category of paddy, it has increased nearly by 4% during the period. 2) Garden has increased by 22.3%.

On the other hand, 3) dry land has dramatically increased from 877.75 acres up 2742.43 acres, 3.12 greater than the figure of 1903 with two peaks around 1960's and 1970's. Marked increase in the extent of dry land in these two periods is mainly due to the successive land-grant of surplus land vesting in the State Government of Karnataka to the former tenants and the landless Scheduled Castes of the Scheduled Tribes.

Most of them once were under close surveillance of landlord. Fig. 5–3 shows how private land had been extended over to the outer fringe, centering on the originally settled areas.

Before 1897, private lands were concentrated on the valley plains and the river terrace where secured supply of water enabled them to grow paddy crops all the

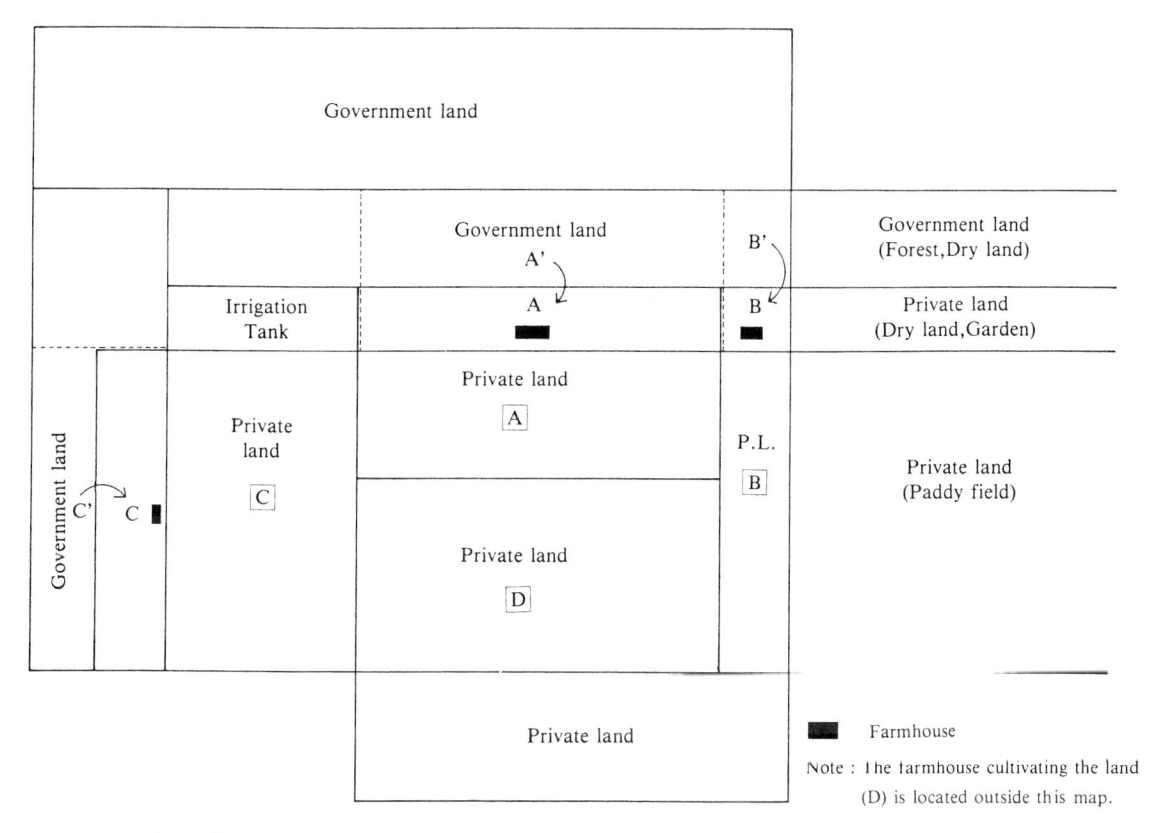

Fig. 5-3 *Kumki Darkas* system in Naravi

year around. When it comes to the next period from 1898 to 1935, private lands were extended outward and filled up the unoccupied spaces among the originally settled areas and then extended further to the outer edge from 1951 to 1969. The remaining unoccupied land is government land. The outer edge was formly held by State Government and was composed of either dry land with bushes and shrubs or forest land.

As clearly seen in these spatial process of expantion of private land, villagers have expanded their private land and taken in successively adjacent areas, most of which belonged to State Government. This processes been caused not only by an increasing demand for paddy garden and others, but also the production of green manure, applying to paddy along with the cow dungs.

The latter factor can not be dismissed through the view point of land grant as well as an ecological balance of man-made ecosystem under the subtropical climate.

In order to make these points clear, it is necessary to give a brief account of traditional common rights to collect green manures from the forest land held by the State Government.

The following example is enough to show how the villagers excercise their customary rights to get green manure. Supposing that there is a vacant land (government land) which is adjacent to the land of cultivator, he has a right to get green manure from there as much as he needs. This common rights have been long observed among the villagers since British ruled over this region.

What is more important is the following point; the common land which is covered with shrubs, baskes and thick forest has often used by the adjacent farmers and eventually appropriated by him in due course. Then, it has been granted by State Government.

This land granted by Government is locally called as "*Kumki Darkas*". *Kumki* means helping or assistance. When it once become private land, it is called as "*Darkas* land". Before granting this land, it is called "*Anadena*", or Not Occupied land.

The main purpose of this land-grant policy is based on the ground that State Government should be in a position to secure a sizable green manures necessary for fertilization of paddy field against constant shortage of organic manures.

Land-grant system as above mentioned has been worked out as a safeguards against the constant shortage of organizers caused by run-offs of various nutriments owing to heavy rains during the monsoon seasons.

We can find out this system in the dispersed villages located on the steeper slope of the Western Ghats, prevailing in especially in South Kanara District.

The system of land-grant can be detailed in Fig. 5-3. We suppose vacant government land atops the diagram to which several private land is adjacent. The farmer A shall gets a big part of government land, because one side of this elongated land lies in close pararell with government land. He gets government land just as long as his land extends. Small share of government land belongs to the former B. For the length of his land bordering government land is shorter than A. The case of C is just same as A and B. But as for the case of D, he is not entitled to government land. Because any side of his land does not touch to the government land. Maximum extent of land, which shall be granted for green manures should not exceed more than 4,5 square chains. One chain is equal to 66 feet in Gunter's Survey and 110 feet in Engineering Survey.

Land-grant system originated in the common rights to get green manure from the unoccupied government land and has been strictly kept on in these ways.

However, there has been growing number of farmers who are very anxious to get more dry land, needed for green manures and grazing land as well as fuel timbers. In response to these recent trends, the State Government has built up traditional land-grant system into Land Reform Act of Karnataka in an unified way. Fig. 3 shows the very process that expansion of private land is extend toward outer edge of the original settlement and now sprawling up to gentle slope of hill.

The government land, which is supposed to be granted for the former tenants, the bonded labourers, has come into their possession after paying the counter value. They are entitled to get financial assistance from the scheduled banks. On the grant of land under section 77 they have the option to deposit money equivalent to the purchase price of land either in a lumpsum or in annual instalment not exceeding twenty years. In the case of annual instalment, annual interest is 5.5%. The amount of these purchase price shall be advanced by State Land Development Bank or from a Credit Agency as defined in the Karnataka Agricultural Credit Operation and Miscellaneous Provision Act, 1974 (Karnataka Act 2 of 1975).

Tenants and agricultural labourers can get finance from the scheduled bank and select a loans out of various types of loans ranging from short term, to long term loans with several repayment systems. But the current bank rate is standing at 15% per annum, 2.73 times more than prescribed rate in the Land Reform Act. Availabilities of money at every level of Banks money positions are so tight that borrowers have to bear high interest rate. The bank rate of Reserve Bank of India is 9%, however, when it comes down to the level of Apex Bank, 2% is added, then at the district level, Central Land Development Bank and District Central Cooperative Bank, another more 2% is added to it. Final lender, primary Land Development Bank marks up again 2%. Total surcharge interest adds up to 6%. Admittedly it is extremely high rate of interest. (Fig. 5–1)

Nevertheless, financial assistance has played a role of critical importance in promoting the redistribution of surplus land vested in the State Government among the landless peoples.

A far-reaching consequences of the land reforms shall crop out in a wider extent from now on. As matters stand an entangled derangement of the rural community seems to have been responsible for hasty enforcement of the radical land reforms.

A premature land reforms fail in vain and leave the source of caramity in the near future. What is most needed is far sighted policy-making in the long run to secure well balanced development of rural community. It is easier to have hindsight than foresight. The radical land reforms have acted undoubtedly as breakthroughs in the arena of rural India and paved the way to sustained economic growth.

However, vicious cycle of poverty is still now working on. It should be borne in mind that there is no instant remedy for this. Radicalism has to be replaced by gradual progressivism.

References

Government of Karnataka (1982): *The Karnataka Land Reforms Act, 1961 and the Karnataka Land Reforms Rules, 1974*, Department of law and parliamentary affairs, Bangalore.

EPILOGUE

In this research monograph six villages were selected out for sample survey to make clear structural changes of rural landscape and salient features in regional dynamics in Wesern Ghats region.

For this purposes, the sample villages Bidarakere and yerdona including Kindi camp were selected from drought prone North Maidan region, on the other hand, Kurbathur Naravi were selected from Malnad and West Coast region, which belongs to tropical monsoon and rain forest zone, in accordance with reconnaissance and traverse survey.

Taking the physiographic, and topographic features in Western Ghats into account, these two regions are sharply contrasted with each other in climatic conditions, soils land use, mode of irrigation system, and development policies as well, as clearly tabulated below. Results are as follows;

1) In the sample survey village, Bidarakere, there has long been observed dry farming system based mainly on tank irrigation with contour bunding. owing to environmental constraints.

Village peoples have traditionally observed time-honoured institutional customs and manners such as those, extended family, dowry and inheritance system, dividing landholdings equally among sons in a way they could survive by themselves in severe natural environment.

This is, in a sense, adaptive responses, still remaining on a sustainable economy stage They, almost of all peoples have long tended to cling to an old fashioned Indian ways of life.

In other wods, they have to cling to traditional cob-web of daily life, which is strictly interwoven into Indianism. Unless otherwise they can not survive. In this sense, main actors on semiarid scene could be termed as *traditionalists*.

2) On the contrary, second sample survey village, Yerdona is favorably situated on the distributary under the command area of Tungabahadra Dam, notwithstanding severe drought prone region.

With the introduction of irrigation, acreage under paddy field have dramatically increased in acreage, and commercial crops became also widely cultivated.This has set spur to mutiple-cropping, and then increased efficiency, productivities of land utilisation, and income level as well. Some of enterprising farmers have not only succeded in expanding their acreages of agricultural holdings, but also gradually taken a lead to regional development, generating thereafter growth poles in this region., which give an infuential impact to neighboring areas.

In this sense, an impotant part of the economic prosperity rests on irrigation rush in this region.

A handful of successful farmers have competing actively with each other and begun to play a roles of vital importance in opening up new era. Paying careful attention to these respects, author categorized this band of group *as Innovationists*.

On the other hand, social mobility through class differentiation among the communities (*jati*), caused by widening income disparities became much more prevarent than ever before.

At the same time In-migration flows gave a decisive blows to the traditional village life, and feelng of belongingness to the same community This phenomena finally resulted in disintegrating social cob-web of rural community, *viz* social unrest in Yerdona

3) In Naravi village, Karnataka Land Reform Act was enacted As the result, traditional absenteeism had been completely uprooted. a handful big farmers were thoroughly wiped out across the almost of all state. Owner-cultivator class was thus created. From institutional view point this land reform has played a role of critical importance, succeded in equalizing landholding from stock level,

Intended objective of this land reform by Karnataka Government should be appropriately estimated in due time. Basically this act was calculated to ease socio-economic gaps the have and the have-not in term of agricultural holdings in a rural society.

Objective of the act was realized to some extent,

but in fact from its very start, class stratification has widely taken place as time goes by. Price mechanism is always working on either in agricultural market, labour market, land market, and money market as well.

As far as this economic mechanism continues, social stratification goes on hand in hand with its inherent powers reshuffling hierarchical clusters.

In some market oriented Villages, it is certainly true of Naravi.

4) In Kurbathur and Yadavarahalii, favourably endowed with both physiographic and agro-climatic conditions, commercial and plantation crops as paddy, coffee, cardamon etc are richly grown, in these sample survey villages and its neighbouring areas.

Among plantation crops coffee and cardamon, arecanuts coffee plantation has long time honoured tradition. At first several British planters began to cultivate coffee trees, then spred gradually its acreage.

After the Independence, Simoga District, especially manjarabad Taluk, in which sample survey villages are included came to gain fame as a one of main coffee producing areas in India, for its quality and quantity as well. In this respect sample survey villages are well endowed with natural environments and abundant in natural local resousces. Staple products as these in this localities have been blessing rural economy up to present..

Notwithstanding capricious onset of monsoon, paddy can be grown at least 2 times a year. Thanks to climatic conditions, production of both extremely labor-intensive paddy and plantation crops are closely related with each other in terms of profits and losses as a counterbalance and need for leveling seasonal allocation of labor around the year.

In each regions six sample survey villages, peoples have accommodate themselves to the traditional way of life and severe natura; settings, in some cases, getting accustomed to changing socio-economic environments, Quick response to changing outer world has been vital point for survival for them. First of all, a few innovationists have succeeded in overcoming their difficulties and finding ways of theirown adaptability should comes first in this context.

In geographical sense, regional characteristics are well reflected on their various ways of economic life in rural scenes. Areal differentiation of economic mode of life in each regions speaks of adaptabilities to natural environments and changing outer world by themselves, landscape, landuse pattern, settlment, and mode of life speaks, and tells its own tales.

In this respect, A series of development policies after Independence, either on national level or State / District, in some cases, Taluk level have to be once again scrutinized minutely, from geographical pont of views, based on yearly and region-wise set of statistical data, which is being summarized as a format, *geomatrix*.

From this point, we will be able to undertake empirical model building for further regional development policy measures. Growth pole theory based on regional science provide us one of strategic view points.

What is now most needed is to accumulate systematic set of geographical data collected from field work for that purpose.

The most reliable and testable data, on various scale either, macro, meso or microscopic scale, obtainable from geographical field work could give us a clue to properly interpret a course of modernizing processes of rural India after Independence.

References

Command Area Development Authority, Tungabhadra Project (1976): *Two decades of development, Tungabhadra Project.* 49p.

Command Area Development Authority, Tungabhadra Project (1976): *Report on re-examination of cropping pattern under TBP, Left Bank Canal.* 116p.

Krishna Water disputes Tribunal (1973): *The report of the Krishna water disputes tribunal with the decision.* Vol.1, 123p.

Singh, Tapeshwar (1978): *Drought prone areas in India.* New Delhi, People's publishing house, 124p.

INDEX

A
absenteeism 97
Agro-climatic region 19
Andra-Pradesh 63
Arabica 66
Arabica variety 65, 66
Arecanuts 78, 82

B
Bidarakere 2
Brief History of Coffee 64

C
canal irrigation system 18
cardamom 61, 67
cashewnuts 78, 82
Circulation of agricultural reproduction 78
coffee 61, 63, 65
Coffee *Arabica* 60, 64
coffee, cardamom 61
commercial crops 19, 28, 37
commercialization 37
common rights 95, 96
Community Project 85
Conjugal family 12
contour-bunding system 1, 97
coolies 47
Crop rotation 29
Crop-combination 20
cropping patterns 17, 19

D
Dani system 85, 86, 87, 90
Dani-Vokkalu system 86, 87
dowry 11, 17
drought-prone area 3
drought-prone region 97
dry farming 2

E
euthanasia of landlord 92

extended or joint family 11, 12, 13

F
family structure 5
family composition 6
family types 7
Former Mysore state 2

G
geomatrix 98
Government of Karnataka 94
Group *Panchayat* Office 54
growth poles v, 97, 98

H
Hassan District 65
high-yielding varieties 19

I
Indian planters 65
individualism 17
inheritance system 9, 14
Innovationists 97
irrigation canal 45
irrigation canal system 2
Irrigation Rush 44, 45, 97

J
Janatha houses 55
Jati 22, 46, 55, 97
Job rush 49

K
Karnataka 63
Karnataka Land Reform Act 57, 78, 97
Kerala 63
Köppen's schema 3
Kumki Darkas 95
Kurubathur and Yadavarahalli 51〜57, 59〜69

L
land tenure system 92, 94

Land-grant system   96
landlordism   87

M

Malnad Villages   51
Monsoon   19, 63, 65, 71～72, 98
Mysore Land Act   85

N

Naravi   71～83, 85～96
New Frontier   44

O

onset of monsoon   18, 68
over-grazing   63

P

Paddy   62
patriarchial inheritance   10
Plantation Crops   63
population pressure   13, 14
profit-oriented crop   19
Profit-oriented farmers   2

Q

Queen of Spices   67

R

Robusta variety   66

S

sample villages   2
Scheduled Castes   89, 94
Scheduled Tribes   89, 94
self-sustained growth   1
Shimoga   65
soil fertility   37
State Land Development Bank   96
subdivision of agricultural holdings   15, 17

T

Tahsildar   89
Tamil Nadu   63
The Karnataka Land Reform Act   85
traditionalists   97
Trewartha's classification   3
Tungabhadra Dam Project   23, 47
Tungabhadra Left Bank Main Canal   45

V

vicious circle of poverty   1
Vokkalu   90

W

West Ghats region   18
Western Ghats and Eastern Ghats   63

Y

Yerdona   2, 24～49
Yerdona camp   46

Regional Dynamics in Modernizing India
Iwao MAIDA

2010年2月25日　第1版第1刷発行

著　者　　米田　巖
　　　　　まいだ　いわお

発行者　　渡辺　政春

発行所　　専修大学出版局
　　　　　〒101-0051　東京都千代田区神田神保町3-8
　　　　　　　　　　㈱専大センチュリー内
　　　　　電話　03-3263-4230代

印　刷
製　本　　藤原印刷株式会社

装　丁　　京尾ひろみ

Ⓒ Iwao MAIDA 2010 Printed in Japan
ISBN 978-4-88125-232-8

本書は，平成21年度専修大学図書刊行助成の制度による
刊行図書である。